DATE DUE

PRINTED IN U.S.A.

MORTAL RITUALS

MORTAL RITUALS

What the Story of the Andes Survivors
Tells Us About Human Evolution

MATT J. ROSSANO

Columbia University Press
New York

Columbia University Press
Publishers Since 1893
New York   Chichester, West Sussex
cup.columbia.edu
Library of Congress Cataloging-in-Publication Data
Rossano, Matthew J.
Mortal rituals : what the story of the Andes survivors tells us about human evolution /
Matt J. Rossano.
pages   cm
Includes bibliographical references and index.
ISBN 978-0-231-16500-6 — ISBN 978-0-231-53546-5 (electronic)
1. Evolutionary psychology. 2. Behavior evolution. 3. Taboo.
4. Airplane crash survival. I. Title.
BF698.95.R67   2013
155.7—dc23
2012039649

Columbia University Press books are printed on permanent and durable acid-free paper.
This book is printed on paper with recycled content.
Printed in the United States of America
c 10 9 8 7 6 5 4 3 2 1

COVER PHOTO: © Popperfoto / Getty Images
COVER DESIGN: Milenda Nan Ok Lee

Contents

Preface

Saturday, September 1, 2012, a few miles west of Independence, Louisiana (about fifty miles northwest of New Orleans). I just had my first shower in four (hot and stormy) days. Hurricane Isaac came through on Tuesday and Wednesday, knocking out the power on Wednesday afternoon. It did not return until earlier today. Since we live in a rural area, no power means not only no lights, television, air condition, and Internet but also no water (the pump on the well goes dead). Three days of "living primitive" were uncomfortable but manageable.

How long could you survive in your house without power or running water? A few days? Maybe a week or more? Being prepared for such a possibility would certainly make a difference. My family and I had the benefit of past experience—Hurricane Andrew in 1992 and the infamous Katrina of 2005. That experience taught us how to prepare the house for a storm and its aftermath. We knew we had to store water—fill the bathtubs, the washing machine, and the ten-gallon containers in the storage closet. We filled the freezer with ice and put it on the coldest setting; bought plenty of dry goods (pop tarts, peanuts, breakfast bars); and checked the flashlights, batteries, candles, charcoal, and lighter fluid. For a while, we were going to be living a bit more like our great-grandparents, so we'd better get ready for it.

Our preparation had to be not just physical but also mental. During the storm, life would be more arduous. Merely using the toilet, for example, would require refilling it after each use. After the necessary thirteen scoops and dumps from the water stored in the tub, our backs would ache, and the day had just started. Life would also be more monotonous. We would be stuck in a dark, hot house with the same unshowered people day after day, with few if any distractions. Get your mind right, I told myself—be patient, don't whine, and be open to the unexpected lessons of your circumstances.

Our great-grandparents did just fine without air-conditioning, running water, and television. In some ways, they were probably happier than we are today. When one's goals are whittled down to just what needs to be done to get through the day, the importance of those activities is suddenly magnified. For example, each evening during our family's post-Isaac existence, the darkening house chased us onto the patio, where to the flickering illumination of citronella torches, we talked, laughed, and listened to the battery-powered radio. It was some of the most enjoyable family time we ever had. My suspicion is that such gatherings were more common in our great-grandparents' time than today. These gatherings may count as what I call in this book an "ancient way," that is, a behavioral pattern more common to our ancestral past, which reemerges when current cultural conditions change.

This book tells the story of a group of young men whose lives were violently reduced to the basics of daily survival. The Andes survivors' story is remarkable and leaves us wondering: How did they do it? This book explores one answer: They did it by accessing the resources of their own human legacy. Their ability to do this, I argue, was enhanced because of a specific type of preparation they brought with them. They were a team, and not just any kind of team. They were a rugby team, and rugby got their minds right when confronting the challenge of survival. Rugby forced them to think in an "ancient way" that they were individually insignificant and that the greater good of the group was paramount, an attitude rather alien to the modern West but essential to our ancestors' success.

In this event we also see something else quite noteworthy: the hazy reflection of human evolution itself. Not that the Andes survivors became a group of prehistoric hunter-gatherers. That would have been impossible. Instead, the actions they took and strategies they employed in their struggle to survive had deep evolutionary histories. In other words, their story has an evolutionary backstory, which is what I attempt to bring to the fore. This backstory makes it a truly *human* story, one that we all share simply because of our common humanity.

For example, the survivors organized themselves into a complex, hierarchical (as opposed to egalitarian) social system. Where do complex social systems come from? They have an evolutionary history that extends back to our primate heritage, through our hunter-gatherer forebears and on into settled agriculture and the first civilizations.

Pushing the questions even deeper, we could ask, "How were they able to form themselves into a complex social system?" The answer here draws on our uniquely human cooperative abilities (abilities that, in this case, were sharpened by their rugby training). If the plane that crashed in the Andes had been filled with cats or camels, they never would have survived. Those species simply do not have the cooperative abilities necessary for forming the kind of complex social system that the Andes survivors created. And why don't cats and camels have those same cooperative abilities? Again, the answer is evolution: those species' evolutionary histories are different from ours'. To understand how the Andes survivors came to possess the cooperative abilities allowing them to form a life-sustaining complex social system, we have to examine our species' unique evolutionary journey.

Undoubtedly, I could have used other, similar events as the evolutionary exemplar. Indeed, the fact that women and children are largely absent from this story excludes the possibility of exploring the evolutionary background of the sexual division of labor. Of course, no single event can encompass all the important evolutionary issues relevant to humans. The advantage of this event is that the relevant evolutionary issues are fairly clear and are embedded in a highly gripping human drama. To put it more bluntly, it's a really good story.

Each chapter of this book opens with a scene based on accounts of the event as they were told in *Alive: The Story of the Andes Survivors*, by Piers Paul Read in 1974, two years after the crash; *Miracle in the Andes: 72 Days on the Mountain and My Long Trek Home*, by Nando Parrado, one of the survivors, in 2006; and interviews with the survivors. These passages introduce a particular issue, such as the decision to eat the dead, the creation of their social order, and the rituals and routines of daily life. I then examine and discuss the scene and the issue it raises before presenting the evolutionary backstory of that issue. With that evolutionary understanding in hand, most of the chapters conclude by returning to the Andes survivors and looking again at the original issue, now from a more informed perspective.

Many people deserve thanks for making this book a reality. First, of course, is my family, whose love and support make possible projects like this. My colleagues in the Thursday science and religion lunch group were an unceasing source of ideas, critical analyses, and challenging viewpoints. They and the students in my graduate seminar in the spring of 2012 added much to this book. I'm also deeply grateful to Piers Paul Read and Nando Parrado for their informative, vivid, and inspiring accounts of the Andes event. Readers of this book are strongly encouraged to read their books as well. Numerous reviewers gave generously of their time, providing helpful and thoughtful critiques of the manuscript as it was being prepared. The final product was immeasurably strengthened through their efforts, and I am in their debt. My thanks go also to Karen Phung and Timothy Lynch for their proofreading. Finally, the editors at Columbia University Press were a delight to work with. Patrick Fitzgerald and Bridget Flannery-McCoy were enthusiastic about the project from the start, and their encouragement, diligence, and expertise were essential to transforming a rough idea into a polished work.

Even though my name is on the front cover, what is best about the book owes its origin to many others.

MORTAL RITUALS

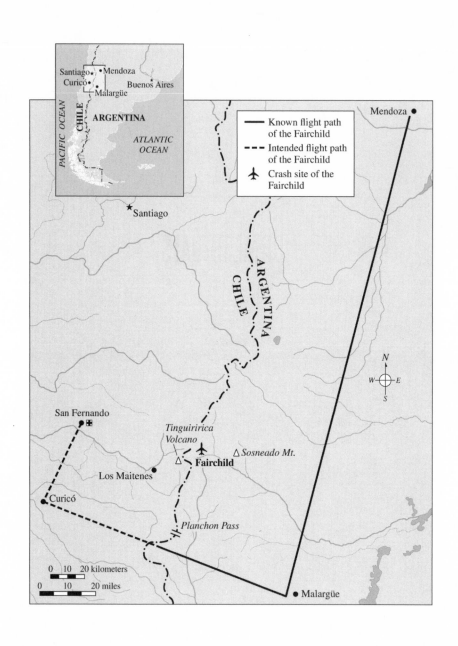

The Crash of Flight UAF 571

We know *what* the pilot did wrong: he badly miscalculated the plane's position. What we don't know (and probably will never know) is *why* it happened. When Colonel Julio Cesar Ferradas[1] radioed air traffic control in Santiago, saying that he had just passed Curico on the western (Chilean) side of the Andes, he was actually some fifty miles east of Curico, deep in the mountainous Planchon Pass. When the controller gave permission for Ferradas's American-built Fairchild F-227 to begin its descent toward Padahuel airport, the plane proceeded to plummet into territory so remote that the peaks were unnamed. Thick clouds obscured the rugged slopes below, and by the time the plane was low enough to visualize the terrain, it was too late.

Possibly the plane was at fault. As a class, the Fairchild F-227 had a poor safety record. In the two years of the planes' production, they tallied twenty-five incidents with nearly four hundred fatalities. This particular plane, however, was nearly brand new, having logged fewer than eight hundred hours in the air, and it was well-equipped with both a VHR Omni directional range and an ADF (automatic directional finder) radio compass.

Possibly the pilot was to blame. Ferradas may not have accounted for the strong headwind buffeting his craft when he made his initial turn into Plachon Pass. But this was his thirtieth Andes crossing; surely, accounting for the wind was by now routine. The experienced air force colonel could hardly be accused of being incautious. The day before, when the weather turned questionable, he scotched his original plan of a direct flight over the mountains from Montevideo to Santiago and instead landed in the Argentinean foothill city of Mendoza to await better conditions. Earlier this very day, he had chosen the more indirect but safer Planchon Pass through the Andes rather than a straight but riskier flight through Juncal. He did this even though another pilot, having just crossed Juncal, told him that while conditions were not good, they were probably no problem for the Fairchild.

The cold facts of Uruguayan Air Force Flight 571 are by now well known: A chartered flight carrying forty-five passengers and crew, most of whom were affiliated with the Old Christians rugby team, crashed in the Andes on October 13, 1972. Seventy-two days later, two ragged, emaciated survivors, Fernando "Nando" Parrado and Roberto Canessa, emerged from the mountains and spotted a Chilean peasant, Sergio Catalan. Catalan alerted authorities, and a subsequent rescue pulled another fourteen young men from the Fairchild's wreckage, eleven thousand feet high in the Andes.

The "Miracle in the Andes" was a crucible punctuated by a series of dramatic, gut-wrenching events: the crash itself, the gruesome decision to eat the dead in order to survive, the avalanche that magnified their suffering and further culled their depleted numbers, the numerous failed expeditions to the outside world, and the final, successful expedition. But this book is not about those events, already so well documented and explored in other venues. Instead, it is about what those events tend to obscure, the equally trying but mundane business of surviving, of hanging on one more day, day after day. What sustained those sixteen survivors for seventy-two days was a combination of many things, among which were teamwork, faith, and a well-organized social system. Behind all these life-sustaining factors was ritual, that is, regimented, purposeful, highly meaningful intentional actions that kept reminding them, day after day, who they were and what their lives were all about.

Mortal Rituals

"To thee do we cry, poor banished children of Eve: to thee do we send up our sighs, mourning and weeping in this vale of tears." This is a verse from the prayer "Hail Holy Queen," which is recited during the rosary, a Catholic ritualized prayer. The nightly rosary was one of the regular rituals the Andes' survivors used to keep up their spirits, to give them hope and strength. For them, the prayer's spiritual banishment had become a physical reality. The "vale of tears" was no longer a metaphorical place east of Eden; it was a glacier (later aptly named the "glacier of tears") perched eleven thousand feet high at the base of an unnamed Andean peak. In this frozen place, ritual took on its greatest urgency.

Ritual is essential to social life. We use it to create increasingly complex social systems so that we can rely on one another for survival. In doing so, we become increasingly insulated from nature's hazards and its beauty. Ritual is also a way of wresting control away from nature. Nature moves the seasons, but ritual marks them, makes them "official" to the human world. We are born, mature, struggle, prosper, and die at nature's command; but it is ritual that welcomes us into, makes us full members of, and dismisses us from the human community— celebrating our successes and lamenting our sorrows along the way. For our ancestors, ritual had mortal consequences. It was ritual that sustained humanity over its long evolutionary journey, just as it was ritual that sustained the Andes survivors.

Ritual's Promise

When Parrado and Canessa found Catalan, the hour was late, and the rushing water of the Rio Azufre made communication nearly impossible. Initially, Catalan thought they were lost tourists, possibly hunters stranded in the mountains. In the shadowy dusk, their dazed, disheveled appearance could easily have led him to entertain more menacing thoughts about these strangers frantically waving at him across the river. Could they be thieves, murderers, revolutionaries on the run from the army? Desperately, Parrado and Canessa sought for a way to make him understand. "For the love of God don't fear us; help us!" In that moment, to save them, they called on the same thing that had sustained them in the mountains: ritual. In an unmistakable act of supplication, Parrado fell to his knees, clasped his hands, and begged for his life. The message was clear but also rich with social irony: A son of the privileged Uruguayan bourgeois demonstrates his dependence on a simple Chilean peasant. Parrado's expression was more than just a claim of helplessness; it was a promise. "I am no threat. I mean no harm—just help me." Against the backdrop of the quickly darkening sky, Catalan returned Parrado's gesture with a promise of his own, shouted above the roaring waters of the Rio Azufre. "Tomorrow," he said.

Long before word could issue promise, gesture was doing so. Or when word was inadequate to ensure promise, gesture could fulfill. Ritual is enacted promise. Ritual promises tomorrow, and the hope of tomorrow sustains.

CHAPTER ONE

Natural Versus Civilized

SCENE 1: THE VIRTUE OF RUGBY

[Nando Parrado describing the brutal action of the rugby scrum] The remarkable thing is this: at the very moment of success you cannot isolate your own individual effort from the effort of the entire scrum. You cannot tell where your strength ends and the efforts of the others begin. In a sense, you no longer exist as an individual human being. For a brief moment you forget yourself. You become part of something larger and more powerful than you yourself could be. . . . No other sport gives you such an intense sense of selflessness and unified purpose. (*Miracle in the Andes*, 16)

Ancient Ways

The players on the Old Christians rugby team first learned their sport from stern Irish monks. The monks arrived in Uruguay in the 1950s to start the Stella Maris College, a boys' school. For the monks at Stella Maris, rugby was not just a sport, it was moral training. Selfless striving for a common goal and individual sacrifice for the good of the team were its essential life lessons.

On the surface, rugby appears to have little in common with a plane crash. But on a deeper level, there is something "un-civilizing" about both. Both entail a detachment from the order, comforts, and social norms that make modern life safe and predictable. Rugby does so intentionally, thereby providing monks with a way to mold adolescent character. A plane crash does so unintentionally. It is a shocking reminder that despite our best efforts, we are unable to control everything: systems fail, humans err, and we are left scrambling to pick up the pieces.

It is a mistake to dichotomize "civilized" and "natural" as if they were polar opposites. By nature, humans want to "civilize" their environments in the sense of making them more amenable to human success and survival. In the course of doing so, their world becomes more predictable, stable, orderly, and, in recent decades, highly technically complex. A by-product of this "civilizing" process is that it can create cultural conditions that contrast with those of our ancestral past, such as imposing more severe behavioral restraints than those our ancestors typically faced.

For example, one way to make things more orderly is to reduce male aggression, and an effective way of doing that is to force everyone to take one lifelong mate (thus reducing the tendency of males to fight over females) or to insist on courts and legal proceedings rather than personal retribution as a means of settling perceived injustices. The reduction in violence benefits everyone, but it also curtails our sexual and competitive appetites in ways largely alien to our evolutionary past. Not surprisingly, easing some of these more recent cultural restrictions often results in a reemergence of ancestral behavioral patterns, a phenomenon that I call a return to "ancient ways."[1]

Are ancient ways more natural for humans? Maybe, but not necessarily. By nature we are designed to seek acceptance and success in our cultures, be they modern or ancient (Richerson & Boyd, 2005). Sometimes, however, the pursuit of that acceptance and success entails conforming to cultural restrictions that inhibit some of our natural inclinations. For example, there's plenty of evidence that by nature, humans are mildly polygamous in their mating habits (Buss, 1994; Daly & Wilson, 1983). That is, we prefer more than just one mate. For males, multiple mates mean more opportunities for reproduction and thus potentially greater reproductive success, compared with strict monogamy. For females, more than one mate can produce more resources for their offspring and, likewise, potentially greater reproductive success.

Traditional societies typically view marriage as important for raising children, although lifelong monogamy is generally not the norm (see Konner, 2010). At the extreme end are the Ache—traditional hunter-gatherers of the Amazon—among whom by age forty the average woman is likely to have been married and divorced nearly a dozen times (Hill & Hurtado, 1996). The Ache are somewhat exceptional, however, as the low male/female sex ratio makes it hard for females to keep one man's attention for very long. The !Kung San, traditional hunter-gatherers of southern Africa, may be more typical. Shostak (1981) examines in detail the life of Nisa, a young woman of the !Kung San. Nisa had two failed marriages as a young teen before a more stable marriage later. Unstable marriages early in life appear to be fairly common among traditional people, with increased stability when offspring are produced (Goodman, Estioko-Griffin, et al., 1985; Konner, 2010, 520–524).

Psychologist Kevin MacDonald's careful analysis (1995) of the rise of monogamy in the West credits centuries of strong influence by the Christian church as instrumental in making monogamy the social norm in Europe (a conclusion reiterated in a more recent study of monogamy by Henrich, Boyd, and Richerson, 2012). But as the influence of religion has waned in the West, divorce and various forms of multiple matings have again become increasingly common. Indeed, a recent survey of Americans found that the proportion of married adults has reached a low point and that married couples soon will be a minority (see http://pewresearch.org/pubs/2147/marriage-newly-weds-record-low). For some people, this reflects the moral collapse of Western civilization, but for evolutionists it is just another example of an "ancient way" reemerging. As cultural restrictions have eased, humans (for better or worse) have settled back into a historically more common (and probably more natural) mating habit—staying together just long enough to satisfy their evolved reproductive interests.

Ancient ways need not always be more natural ways. For example, nearly all our anthropoid (monkey and ape) relatives live in some form of social hierarchy. Thus, we might argue that social hierarchy is natural to humans. Traditional hunter-gatherers, however, live in aggressively egalitarian societies in which any attempt at individual social dominance is met with derision, ridicule, and sometimes violence. So when French revolutionaries dispatched the old aristocratic regime—declaring liberty, equality, and fraternity for all—they may have been returning to an ancient (hunter-gatherer) egalitarianism, but whether it was a more natural (primate) social order is questionable.

A number of factors permit the reemergence of ancient ways. The decline of monogamy highlights one factor: the weakening of religious and/or cultural restraining forces, which can be abetted by another: technology. Culturally imposed monogamy becomes increasingly difficult to sustain when reliable contraception allows men and women to follow their natural sexual inclinations while avoiding the complicating issue of progeny. A third factor is the breakdown of powerful social and political forces. Great empires often imposed peace on long-standing, parochial, tribal conflicts within their borders. When empires fall, these conflicts often reemerge after lying dormant for decades or even centuries. The conflicts themselves suggest that tribal or group-based competition was common in our evolutionary past, especially when scarce resources were being contested (Ember, 1978; Kelly, 1985; Soltis, Boyd, & Richerson, 1995). More than this, however, the way that local conflicts play out also echoes the ancestral past. The sight of teenage boys carrying weapons in some remote jungle of Africa, Southeast Asia, or Latin America is often particularly unsettling to Westerners. But for our ancestors, boys with weapons were probably not so uncommon. Long before there were organized armies and regimented military training procedures, teenage boys eager to prove their manhood were the main source of a tribe's warriors.

Finally, ancient ways can arise when social institutions are absent or ineffective. For example, during much of the eighteenth and nineteenth centuries in the American South and West, law enforcement and legal institutions were unreliable, so social control was based on an honor system or "honor culture" (Cohen & Vandello, 2001). In this system, males sought to establish a reputation for both gentility (on the one hand) and ferociousness in defense of personal and familial honor (on the other). Gentility made them attractive to high-status females while at the same time opaque to their male competitors. A fierce and often violent defense of honor signaled to other males a willingness to retaliate unmercifully against those threatening their status or resources. Indeed, a ferocious defense of reputation, status, and resources is a time-honored strategy for enhancing males' reproductive success. Thus it is not surprising that when social institutions for ensuring contracts, property rights, or civil order in general are weak, males revert to a more primal means of guarding their interests: a thin veneer covering a lethal competitive drive.

The Crash and Ancient Ways

[On the mountain] you get closer to the idea of . . . death and you think you are just passing through life, and that life is an accident in which the only real thing is that you're going to die. . . . You were there and saw a friend dead, a friend who ten minutes earlier was alive. (Roberto Canessa, interview, 2002, http://www.viven.com .uy/571/eng/EntCanessa012002.asp)

The crash of UAF flight 571 provided the physical context for the re-emergence of some of these ancient ways. The survivors were violently disconnected from the modern world and all its cultural and institutional supports, and it took some time for them to absorb the full reality of their situation. For a while, they clung to "civilized" ways of thinking and behaving. Steadily, however, they shed their former selves and adjusted to the grim, brutally capricious rules of their mountain home. In a number of important ways, their "new" reality became much more akin to that of our prehistoric ancestors. One obvious similarity was the constant shadow of death.

The Hiwi are traditional hunter-gatherers living in the jungles of Colombia and Venezuela. Before contact with the outside world, the Hiwis' life expectancy at birth was only twenty-seven years (Hill, Hurtado, & Walker, 2007). However, as is true for most traditional societies, this number is skewed by high infant and juvenile mortality rates. Among the Hiwi, once a child reached age fifteen, he or she could expect to live, on average, another thirty-one years. These numbers fall in line with hunter-gatherers generally, whose average life expectancy ranges from twenty-one to thirty-seven and whose infant and juvenile mortality rates are high (Gurven & Kaplan, 2007). More than 40 percent of infants born in traditional hunter-gatherer societies will not reach their fifteenth birthday, although 64 percent of fifteen-year-olds can expect to live to age forty-five. Infections and gastrointestinal disorders account for about 70 percent of deaths among traditional hunter-gatherers, and violence and accidents are responsible for about another 20 percent (Gurven & Kaplan, 2007). Mothers dying in childbirth is also much more common than in modern Western societies. For the Hiwi, about one in every sixty pregnancies is fatal.

The extent to which these trends reflect our evolutionary past is debatable. Some studies argue that our hominin ancestors' life expectancy was very likely shorter than that of present-day hunter-gatherers (Cutler, 1975; Weiss, 1981). While this claim has drawn criticism, there is evidence showing that the presence of an adult "older generation" (thirty years and older) was not a stable fixture in hominin evolution until very recently. Using dental samples from four different categories of hominins (Australopiths, early *Homo*, Neanderthals, and Upper Paleolithic modern humans), Caspari and Lee (2004) found that not until the Upper Paleolithic (35,000 years before present, YBP) did older adults actually outnumber their younger counterparts in the typical hominin social group.

Practically speaking, what does all this mean? Put bluntly, if you or I had been born 100,000 years ago, the odds are that by our fifteenth birthday (if we were among the lucky 60 percent to make it that far), at least one of our siblings would be dead, along with a parent or some other adult(s) in our band. We already would be keenly aware of the unpredictable, ever present specter of death. Forty-five passengers boarded the Fairchild in Mendoza on Friday, October, 13, 1972. The crash itself took twelve lives, and five more were dead by the next morning. The avalanche that struck the survivors on October 29 killed eight more. These punctuated "death events" were preceded and followed by a steady attrition of the injured who could not get proper medical attention, the weak and weary who simply could not hang on, and the disturbing deaths that seem to happen for no good reason at all.

One of those last instances applied to a boy named Numa Turcatti. Numa was originally chosen to be an "expeditionary," one of those who would attempt to climb out of the mountains in order to bring help. This was a testament to his strength and determination. Before the expedition was to depart, however, someone stepped on Numa's leg, bruising it. The injury seemed minor. But it went septic, making it difficult for him to walk and thus making it impossible for him to climb. Once excluded from the expeditionaries, Numa became increasingly despondent. His appetite waned. He grew steadily weaker and, despite the admonitions of his companions, was unable to break the fatal cycle into which he had fallen. Numa died on December 11, just before the final expedition departed. No one dies of a bruised leg, especially not the youngest and strongest among us. That is, no one *today* dies of a bruised leg. But in our past, no injury could be casually dismissed. Any tiny sliver though which death might creep was cause for concern.

The crash plunged the survivors into a more primitive world. But they were not entirely unprepared for this world. Most of them were rugby players, and rugby itself is a training ground for more ancestral ways of thinking.

Rugby and Ancient Ways

We might see rugby's relationship to ancient ways as simply encouraging greater male aggression. Rugby can be violent. Indeed, its early popularity was partly because it was seen as a counterforce to the increasing "femininization" of men resulting from sedentary, urban lifestyles (Nauright & Chandler, 1996). But this is not rugby's most crucial connection to an ancient way. Instead, it is rugby's requirement of a complete smothering of ego in the course of directing one's aggression in concert with others against a common foe. More so than other sports, rugby requires complete submersion into the team. Many sports and activities serve as constructive outlets for male aggression; if this were rugby's only virtue, it probably never would have become as internationally popular as it is today. But rugby is more than just a good way for young men to beat up one another instead of us. It trains the soul as well as the body.

Rugby's origins can be traced to the hard-scrabble English villages of the early nineteenth century. It was here that roughnecks plunged energetically into loosely organized football matches unencumbered by rules or restrictions (Dunning & Sheard, 1979). By the mid-nineteenth century, rugby's popularity was growing in England and in its colonial possessions, where it enjoyed a particularly devoted following. For some, this was cause for alarm, as the violence inherent in the game was thought to encourage hooliganism, vulgarity, and other antisocial traits.

But English schoolmasters saw the game as an opportunity to mold a new ideal of manliness. They and other observers of English society were growing increasingly concerned about the "feminizing" effect of urban life. Disappearing, to their dismay, was the old aristocratic ideal of the military gentleman: a disciplined, brave, Christian man of authoritative yet civil bearing. In its place was something pitifully tame: a sophisticated, cynical urbanite reflexively recoiling from any taint of sweat or dirt. The empire, they feared, was imperiled by the decadent softness of English men. If properly structured, rugby could be a useful tool in reversing this dangerous decay. It gained enthusiastic and articulate advocates who sought to corral the worst of its on-field mayhem while expounding its virtues to a skeptical public (Nauuright & Chandler, 1996). These advocates popularized rugby as "muscular Christianity," whose physical challenge would help boys develop the moral fiber necessary to live honorable, upstanding lives. Rugby, they argued, was especially effective at inculcating three important virtues:

1. Hard work and determination: Life was a tough, competitive struggle, and the ability to endure and persist was essential to success. Rugby required these same traits.

2. Selfless individual sacrifice for the greater good of the group or team. The nineteenth-century newspaper man and rugby enthusiast H. J. Wynyard put it this way: "Football [rugby] is a game which . . . requires that every player of the team shall sink his individuality and, like a part of a machine, work in concert with the other parts." The collective physicality of the scrum, the necessity of passing the ball to advance, and the need to protect teammates and be protected by them hammered home to each player the insignificance of the individual in the pursuit of a common goal. In contrast to other sports, the superstar was a rarity in rugby because no one succeeded by being the center of attention.

3. Self-discipline and the ability to repress selfish desires for the greater moral good. The true man is a servant not to his base desires but to his family, community, country, and God. Rugby develops the self-denying character necessary to embrace the responsibilities and burdens of manhood with the same grit and spirit required for challenging one's foes on the muddy rugger's field. (Phillips, 1996, 82–85; quotation, 83)

It was this selfless mental state that harked back to a more ancestral way of thinking. The notion of individual autonomy is a recent development in our history, and regarding it as a virtue is even more novel and not universally embraced. Eastern and African cultures continue to be far more collectivistic in orientation compared with the West.

"I am because we are . . ."

Our primate heritage makes us naturally social. But even among primates, humans are hypersocial. Over the course of evolution, our hominin ancestors lost nearly all the *individual* traits needed for survival. Gone were the menacingly large canine teeth that chimpanzees use to threaten rivals and hunt prey. Gone were the long forelimbs and grasping feet necessary to climb trees to find fruit and to escape predators. Gone was the hair that protected the naked skin from biting and stinging insects and that allowed infants to cling while their mother foraged. By the time of *Homo erectus,* our ancestors were naked, oddball primates whose fate depended on their smarts, their tools, and one another. Left on his or her own, *Homo* had no chance. Inclusion in the group was life; separation from the group was a death sentence.

The dependence of the individual on the group is reflected in the traditional African proverb "I am because we are, and since we are, therefore I am" (Mbiti, 1970, 141). In traditional African culture, the community is the defining social reality, so individuals find personhood in relation to their communities. Based on her work with northern Sudanese tribesman, anthropologist Janice Boddy reports that the "northern Sudanese with whom I worked do not see themselves as unique entities, wholly distinct from others of their group. . . . Personhood in northern Sudan is relational rather than individualistic; a person is linked corporeally and morally to kin" (Boddy, 2010, 115).

Modern Western philosophers have often searched for the defining aspect of personhood in some quality or trait unique to and universal among humans, such as rationality, freedom of the will, or an ethical sense (e.g., Rawls, 1971; Sartre, 1956). Traditional African philosophy defines personhood not as a biological endowment but as a state achieved through increased incorporation into a community (Menkiti, 1984). This state entails the assumption of roles and responsibilities within communal life and is typically bestowed through a ritual. Thus, personhood in the traditional African sense is something that one could fail to attain or might attain in only a limited degree depending on his or her status or standing in the community.

Ancient literature echoes a similar theme. For example, when we first encounter Odysseus in the *Odyssey* (book V), he is "sitting on the shore, his eyes as ever wet with tears, life's sweetness ebbing from him in longing for his home." The shore is that of Calypso's island where he has been held for some time. We are told that he spends his days sitting "among rocks or sand, tormenting himself with tears, groans and anguish, gazing with wet eyes at the restless sea." A far cry from the cunning hero of the Trojan War. But literary scholars tell us that Homer's depiction of Odysseus is more than just that of a weary wonderer longing for home (e.g., see Vernant, 1996). What we see is a man who has lost his very soul, a man who has become something less than human.

As Odysseus stares out from Calypso's island at the distant smoke fires of Ithaca—his homeland—he realizes that his very humanity has been forfeited, not because he has lost his rationality or his free will or his ethical sense, but instead because he has lost his community.

Seeing Ithaca so close and yet unattainable reminds him of what he is *not*—he is not a king, not a warrior, and not a father or husband. These social roles exist solely within his homeland's communal context. Cut off from that community, he is nothing but a lone, isolated man with no identity, no personhood, nothing.

Homer was part of a largely oral tradition of literature. By their very nature, oral traditions are interpersonal and public. Thoughts, ideas, and stories are shared and discussed, their meaning and significance subject to an ongoing communitywide assessment (Barnes, 2000, 83). This public shaping of ideas includes ideas about the self. Who we are is open to discussion and is inextricably embedded in the tribe, its culture, and its traditions. But widespread literacy encourages a privatization of thought (Goody & Watt, 1963; Ong, 1982). With written language, we can far more readily reflect, evaluate, and reason away from others' input or contradiction. Our sense of self can increasingly be shaped through *intra*personal interaction with our own (written) thoughts rather than through *inter*personal interaction. Widespread literacy, of course, is a very recent development in human history. Thus, from a variety of directions it is clear that the idea of the individual as a unique entity is very recent. Rugby rejects "modern individualism" and returns its players to the more ancestral state of "person-as-collectively-defined."

The Evolution of Teamwork

The hackneyed phrase there's no "I" in "team" does have some truth to it. Teamwork works best when individual egos are suppressed, and humans are specially adapted to be good team players. The fact that the survivors of UAF flight 571 were humans gave them a chance because they could cooperatively organize themselves better than any other species. The fact that most were rugby players, and therefore already well trained in thinking and functioning as a team, gave them an even better chance. Being rugby players helped them actualize and expand a natural human teamwork potential.

Cooperation Among Chimpanzees

As our closest primate relatives, chimpanzees give us a rough idea of what our earliest hominin ancestors may have been like. Chimpanzees have some cooperative abilities. Probably the most dramatic and debated instances of chimpanzee cooperation are found in the hunting behavior of male chimpanzees in the Tai Forest of West Africa (Boesch & Boesch, 1989). Here chimpanzees have been observed working as a team (apparently) to capture and kill colobus monkeys.

The action begins when a group of chimpanzees (usually two to six) spot a potential prey in the nearby trees. The spotter emits a characteristic "hunting bark" and races up the tree after the monkey. Other chimpanzees join in the chase, running along the ground and climbing adjacent trees in what looks to be a coordinated effort to block the monkey's escape routes. Behaviorally, the presence of teamwork seems inescapable, but the interesting question for researchers has been the extent to which these behaviors are driven by an understanding of the complementary roles needed to achieve a common goal. In other words, when one chimp sees another race up a tree after a monkey, does he think, "OK, Ed's got that covered; now I need to cut him off at the pass." Or is it possible that what looks like real teamwork is just an illusion. Maybe everyone is just out for himself. If one chimp sees another chasing a monkey up a tree, he may simply realize (based on past experience) that the best way for him to get the monkey is not to go up the same tree but to go up another one nearby.

To untangle these alternatives, researchers have devised a number of controlled laboratory studies looking at chimpanzees' cooperation. An initial question is whether chimpanzees are capable of understanding another's intentions and goals. This is important for teamwork because we first need to understand *what* another is trying to achieve before we can help him or her achieve it. For example, if I see that you intend to haul a big box through the front door, then I can help you by holding open the door while you bring it in. On this point, chimps generally succeed. For example, if a chimpanzee sees a human attempting to grasp an out-of-reach object, the chimpanzee will help out by getting the object for her (Warneken & Tomasello, 2006). Chimps don't help only humans; they help one another as well. In one study, for example, one chimp watched another trying (but failing) to open a door. The observing chimp knew from past experience that the door could be opened only by removing a pin. The observer went to the door, removed the pin, and opened it for the other chimp (Warneken, Hare, Melis, Hanus, & Tomasello, 2007). In both these examples, the chimps recognized the other's goal and helped him achieve it.

But true teamwork requires more than just helping another achieve a goal. A second critical element is understanding the complementary roles in achieving a goal. So if you already have picked up the box, I need to recognize that a different action (opening the door) is required to get it in the house. By understanding complementary actions, a person should be able to engage in role reversal—that is, if you open the door, then I should immediately recognize that I'm supposed to pick up the box. On this point, chimpanzees fail.

Psychologists Michael Tomasello and Melinda Carpenter conducted a series of experiments addressing chimpanzees' cooperative abilities (Tomasello & Carpenter, 2005). In one, they assessed the role-reversal abilities of three human-raised chimpanzees (note that human-raised chimpanzees often exceed wild or captive mother-reared chimpanzees in social/cognitive skills). In the task, the goal was to get a toy into its proper place, for example, getting a lego block into a toy cup or putting Winnie-the-Pooh into his wagon. The experimenter showed the chimp how the two objects went together and then had the chimp hold the toy (the lego or Winnie) while the experimenter held out the "home" (toy cup or wagon). The experimenter then waited for the chimp to place toy in the home. In some instances, the chimp needed encouragement before finally placing the toy in the home, but eventually he or she got it. Then came the test of role reversal. The experimenter took the object and gave the chimp the home. The experimenter then waited for the chimp to hold out the home *while looking at the experimenter* in anticipation of the experimenter's placing the toy in the home. The experimenter waited in vain. While occasionally the chimp would hold out the home, no chimp ever looked at the experimenter indicating that he or she expected the experimenter to execute the complementary action (putting the toy in the home). This was in sharp contrast to a similar study using twelve- to eighteen-month-old human infants (Carpenter, Tomasello, & Striano, 2005). In this study, even the youngest infants held out the home while looking at the experimenter, obviously anticipating that the experimenter would put the toy in the home.

Tomasello and Carpenter (2005) argued that "looking while holding out" was the critical criterion by which role reversal should be judged. The simple act of holding out the home could be carried out based merely on imitation. Looking while holding out indicated the expectation of a complementary act from one's partner. Thus they concluded that there is a critical difference in the collaborative abilities of young children and chimpanzees. Children understand a collaborative task from a "bird's-eye" perspective. They understand how their role in the task complements and coordinates with another's in order to achieve a mutual goal. By contrast, chimpanzees understand a collaborative task from a first-person perspective. That is, they understand that they are doing X while another is doing Y. Under some circumstances, they can imitate Y, but they do not understand the complementary nature of X and Y in achieving a common goal.

Another study (Warneken, Chen, & Tomasello, 2006) not only confirms this interpretation, but it also demonstrates a critical motivational difference between humans and chimpanzees: It's not just that humans understand joint endeavors better, it's that we *really want* to do stuff together and that we get annoyed when others won't "play" with us. In Warneken and colleagues' study, the researchers had a young child or a chimpanzee join with a human adult in a collaborative task or social game (such as a trampoline game in which players hold different ends of a large fabric and attempt to keep a ball from rolling off). Once again the children (as young as eighteen months) proved to be more competent team players than chimpanzees. Moreover, at one point the adult with whom the child or chimp was partnered stopped performing her part of the task or game. The children were very insistent in trying to get the adult to reengage so that the task or game could continue. Indeed, another study has shown that children even try to get a partner to reengage when the task can easily be completed by the child alone (Graefenhain, Behne, Carpenter, & Tomasello, 2009). By contrast, chimpanzees never made any attempt to get the adult partner to reengage. Thus, in explaining the evolution of human teamwork, we must account for not only the greater skill that humans have accrued over time but also the strong emotional desire to be part of a team.

The Social Norms of Teamwork

So far we have seen that human teamwork involves understanding goals and intentions (an ability shared with our closest primate relatives). Furthermore, it requires understanding complementary roles and being motivated to play those roles (something largely unique to humans). A final important factor is following social norms. In this context, norms refer to morally significant rules of social conduct (for a full discussion, see Rossano, 2012). For example, repaying favors, respecting elders, and showing proper decorum at ceremonies such as funerals or formal banquets are usually thought of as norms, because the failure to follow them harms one's reputation and social standing.

By contrast, a social convention is a less consequential social rule that simply allows for smooth coordination, such as keeping to the right when walking on a crowded sidewalk.

One reason why humans are such motivated team members is because of the strong emotional investment we make in following norms. We love rules, and being a good team member is an effective way of showing others how much we love rules. From a very early age, children seek out the social rules by which they should conduct themselves, and once having acquired those rules, they vigorously enforce them on others.

How do children identify social rules? The answer seems to be that they look for indicators of intentional actions. For example, in one study (Schmidt, Rakoczy, & Tomasello, 2011), three-year-olds watched an adult as she performed certain actions on objects, such as pushing building blocks across a board using another object. In one condition, the adult performed the actions with gestures and nonlinguistic expressions indicative of confident familiarity (signaling that the actions were the "right" ones for the given objects). In another condition, the adult's gestures and expressions indicated that the actions were improvised with uncertainty "on the spot." Later, the children were given the opportunity to interact with the same objects as the adult model used.

Those children who watched a confident, intentional model differed in two important ways from the children who watched an unsure, improvising model: (1) they more often imitated the model's actions, and (2) even more significantly, they more often protested when they saw another "improperly" interacting with the objects. These findings suggest that when children see an adult intentionally doing X with Y; they extract a behavioral rule: "When you have Y, you are supposed to do X with it." This leads to a peculiar phenomenon among young children called "overimitation," that is, imitating intentional actions that seem to have no clear purpose.

Children are quite willing to imitate odd, causally irrelevant actions if adults obviously and intentionally demonstrate those actions. For example, suppose that an adult very deliberately takes a feather and strokes it across the side of a jar before opening the jar to retrieve an object. Having just witnessed this, three- to five-year-olds are then given the opportunity to retrieve an item from the same jar. Lyons, Young, & Keil (2007) found that children imitate the feather-stroking behavior before removing the lid. Similarly, Nielsen, and Tomaselli (2010) found that two- to thirteen-year-olds imitated either rotating a stick three times over the top of a box or wiping a stick three times from front to back across a box before opening it. They persisted in imitating these actions even after one group of children was allowed to discover how to open the boxes directly by hand (without using a stick).

Thus, understanding that the stick motions were not *physically causally* necessary to open the boxes did not deter the children from reproducing the stick-rotating or -wiping actions. Children don't have to know *why* they are performing an action, they simply have to be convinced that they *should* perform the action, and seeing an adult intentionally performing the action is convincing (Kenward, Karlsson, & Persson, 2010). Nielsen and Tomaselli's study is notable also because some of children were Kalahari Bushman, demonstrating the culturally robust nature of overimitation.

Does this imitation truly mean that the children have learned a rule? One reason to suspect so is that children do not just imitate the intentionally modeled action; they enforce this action on others. Indeed, children can be quite vigorous rule enforcers. For example, Rakoczy, Werneken, and Tomasello (2008) had two- and three-year-olds play a game in which two behaviors were demonstrated, each of which achieved the game's goal of getting an object into a hole. However, one of the actions was clearly labeled as part of the game, and the other was not. Later, if another child tried to achieve the game's goal using the behavior that was not labeled as part of the game, the children spontaneously and strongly protested, producing both normative protests ("No! It doesn't go like that") and indicative protests ("No! Don't do it that way"). Interestingly, children did not protest if before engaging in the rule-violating behavior, the player made it clear that she was not intending to play the game but instead simply wanted to "do something" with the game materials. This shows that children not only were highly sensitive to the normative structure of the game but also understood when that structure was applicable and when it was not. Other studies have produced similar results (Rakoczy, 2008; Rakoczy, Brosche, Warneken, & Tomasello, 2009).

Do chimpanzees follow social norms? Some studies (Brosnan & de Waal 2003; Brosnan, Shiff, & de Waal, 2005) have suggested that they might, but a closer examination casts doubt on this. For example, a chimpanzee is more likely to reject a low-quality food reward (cucumber) if he or she just witnessed another receiving a high-quality food reward (a grape). One interpretation is that this rejection is motivated by a norm of fairness: the chimpanzee makes a social comparison of what another received relative to himself or herself. But simply showing the chimpanzee a high-quality reward (without its being offered to another) led to the same rejection, suggesting that a food comparison, not a social comparison, accounts for it (Brauer, Call, & Tomasello, 2006).

Another way to test for following norms is to use something called the "ultimatum game." In the ultimatum game, one player (the proposer) is given a sum of money that he or she is allowed to divide in any way desired with a second player (the responder). Thus, if the proposer is given $10, he or she can keep all the money, give it all to the responder (which no one ever does), split it equally ($5 for the proposer, $5 for the responder), or anything in between. The responder can either accept or reject the proposer's offer. If the responder accepts, both keep their money. If the responder rejects, neither player keeps any money. In terms of pure self-interest, the proposer should divide the money in a way that maximizes his or her gain (e.g., $9 for the proposer, $1 for the responder), while the responder should accept any nonzero amount (since acceptance always leads to greater monetary gain than rejection).

The critical finding is that even in anonymous one-off interactions in which the players would not be expected to have any concerns about "paybacks" or social evaluations, both proposers and responders show unexpectedly high levels of cooperation (Carpenter, Burks, & Verhoogen, 2005; Fehr & Fischbacher, 2003; Forsythe, Horowitz, Savin, & Sefton, 1994; Marwell & Ames, 1981; Roth, Prasnikar, Okuno-Fujiwara, & Zamir, 1991). Proposers tend to offer 30 to 45 percent of the money, and responders tend to reject offers lower than this. This suggests that both parties operate using a norm of fairness and are willing to forgo selfish gains in order to adhere to this norm. Furthermore, cross-cultural studies show that the average splits offered by proposers and the acceptance/rejection thresholds of responders show a predictable variation that depends on their culture's norm of fairness (Henrich, Boyd, Bowles, et al., 2001; Henrich, McElreath, Barr, et al., 2006).

Chimpanzees have also been tested using a species-adapted form of the ultimatum game (Jensen, Call, & Tomasello, 2007). In the chimpanzee version of the game, the proposer was shown food items divided between two trays, for example, eight raisins in one tray and two in the other. The proposer made an "offer" by pulling a rope, which moved the trays halfway toward both the proposer and the responder. The responder could "accept" the offer by pulling on a rod that came into reach as a result of the "offer." By pulling the rod, both trays were drawn the rest of the way toward both apes and thus within reaching distance. Failure to pull the rod was consider a "rejection."

Human responders consistently reject unfair offers, so knowing this, human proposers rarely make them. Not so with chimpanzees. Chimpanzee proposers consistently made unfair offers (such as eight raisins for the proposer, two for the responder), and chimpanzee responders accepted nearly all nonzero offers. This finding supported the hypothesis that chimpanzees operate according to a selfish rational maximizing principle rather than by a norm of fairness.[2]

Effective teamwork requires following norms, a set of (often unwritten) rules by which teams operate. Violating these rules usually causes a team member to be alienated from the group and can lead to division and dissention. For example, if you are the team captain or leader, you should always keep your head, avoid excessive displays of emotion, and, no matter how angry you might be at a team member, do not berate him publicly but take him aside privately. If you're a rookie on the team, you should avoid too much talk and instead listen and learn. Finally, everyone on the team must take his responsibilities seriously—others are counting on that. *Homo sapiens* is the only species that recognizes the existence and importance of social rules like these, and we appear to be highly motivated rule followers. Often our scrupulous rule following is a way of building our reputation and gaining status and respect among others, but it can also be done simply out of love for the group.

An Evolutionary Scenario

How did we become such good team players? How did we evolve from a chimpanzee-like level of cooperative ability to our present state? Based on his extensive research comparing human and nonhuman primate cooperative abilities, Michael Tomasello (2011) offers the following three-step evolutionary scenario: First, our hominin ancestors became *obligate cooperative foragers*. This means that to survive, we had to work together when hunting or gathering food resources. Exactly what produced this change is not entirely clear. Physical changes such as the loss of powerful canine teeth or increasing hairlessness (which meant that offspring could not cling to their mother but had to be carried), or increasingly dependent offspring all may have played a role. A mother carrying a baby has a hard time being an effective forager, but two mothers can be effective working together, with one watching the kids and the other gathering berries. A father can feed a mate and baby more effectively if he can bring down larger prey, but he can do this only with the help of a hunting partner or two.

In addition to foraging, a case can also be made for cooperating in the birthing process. One study (DeSilva, 2010) provides evidence that by as early as three million years before present (3 MYBP), our ancestors were giving birth to significantly larger babies in terms of both head size and overall body size. The size of these infants was probably at or near their mother's pelvic opening capacity (Tague & Lovejoy, 1986), which very likely necessitated some type of assistance in birth. Thus, two activities directly related to survival and reproduction (food procurement and giving birth) were probably group activities rather early in our evolutionary history.

The second step in the evolutionary scenario was *self-domestication*. Those hominins best able to work together would have had an advantage over more antisocial types that tried to go it alone. Those who worked together the best were more tolerant of others, were able to recognize common goals and coordinate activities to achieve those goals, and were more motivated to be part of a team—in short, those with the mental and emotional traits that made them more sociable.

Finally, the third step was *cultural group competition* (Richerson & Boyd, 2005; Sterelny, 1996). Different human groups with different cultural norms competed with one another, and the winners were usually those with more effective norms for promoting within-group cooperation. For example, suppose that two groups came into conflict over a scarce resource such as water or a prime piece of hunting ground. Group 1 has a tradition of caring for orphans and widows, but group 2 does not. The males in group 1 may be more likely to sacrifice themselves in order to secure the scarce resource, since they know that their wives and children will be cared for if they are killed in the effort. This could give group 1 a decisive advantage, making it likely that the "care of widow and orphan" norm would spread throughout the human population over time.

Ego Versus the Team

Even though no species on earth cooperates as effectively as *Homo sapiens*, our cooperative tendencies live in tension with our self-interests. We evolved to be creatures highly skilled at using teamwork and cooperation as a means of promoting self-interest. Ultimately, the primary interest of individual humans, as it is for all organisms, is promoting their genetic fitness. We want to be on the team because we expect to get something from it (increased status, team-based rewards that cannot be achieved individually, spoils of victory, etc.). We don't want to be on just any team, we want to be on a winning team. We are designed to guard against sacrificing more for the team than the reward we will gain from it. These "safeguards" against perceived exploitation come in various forms.

For example, humans have a highly sensitive cheater-detection mechanism that is specifically designed for situations involving reciprocal social agreements (Cosmides, 1989). Imagine the following logic problem: four cards with a letter on one side and a number on the other are placed on a table, as shown below:

You are to determine whether the rule "If there is a vowel on one side, then there is an even number on the other" is being followed. You are allowed to turn over two cards. Which two do you turn over? When this experiment was done many decades ago (Wason & Evans, 1975), most people got it wrong. They wanted to turn over the E (to verify that there was an even number on the other side) and the 4 (to verify that there was vowel on the other side). The E selection is all right, but the 4 selection is a logical error. Since the rule tells you only what is supposed to happen when there is a vowel on one side (there is to be an even number on the other), you don't know what will happen if there is a consonant on one side. Thus, a consonant on one side can lead to an even number on the other. So if you turn over the 4 and find a consonant, this does not necessarily mean that the rule has been violated. In other words, the 4 card is irrelevant to evaluating the rule. The other correct selection is the last one. If you find a vowel behind the 7, then you can be certain that rule is not being followed. The fact that most people failed the test suggests that we are not very good logical thinkers. But suppose this problem is recast as a reciprocal social agreement.

Your teenage daughter has been nagging you for weeks to allow her to have a party at the house. Finally you relent, but with the understanding that there will be no underage drinking at the party. In this instance, you have entered into a cooperative social agreement with your daughter: "I agree to let you use the house, but you must agree to allow no underage drinking." You return the night of the party to find the place reeking of beer. Now you want to find out if she has broken her agreement. This situation can be set up as a card problem similar to the even number/vowel situation described earlier. You encounter four teenagers about whom you have information on either their age or what they are drinking. Which two do you want to "turn over" (i.e., get more information about) in order to determine whether the rule "If the person is under twenty-one, then he or she must be drinking a nonalcoholic beverage" is being followed? The teenagers are shown below (in this test, rather than thinking too much about it, just use your gut):

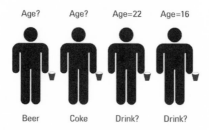

Age? Age? Age=22 Age=16

Beer Coke Drink? Drink?

When people were given the test in this format, most of them made the correct selections: the first one (How old is the person who is drinking beer?) and the last one (What is the underage sixteen-year-old drinking?). For most people, this version was simple. Why is there a difference between the two experiments? Evolutionary psychologists believe that our capacity for logical thinking did not evolve as *general purpose* logical reasoning, but as *social reasoning.* In our ancestral past, we did not need to resolve abstract logical dilemmas, but we did have to resolve practical social problems—"Can Thor be trusted with my spear if I lend it to him?" "If I offer to watch the children, will Hilda share equally with me all the berries she collects?"

For thousands of generations, our ancestors had to work together to survive, but at the same time they had to make sure that their partners were trustworthy. They had to guard against being exploited. A cold evolutionary calculus was at work here. Yes, those who cooperated with others had an advantage over the loners. But those who could persuade their cooperative partners to take the greater risk, make the greater sacrifice, or be satisfied with a smaller reward had an even greater advantage. Our ancestors were the cooperators who were exquisitely sensitive to the "cheats" and "suckers" in their groups.

Another way we guard our interests in social contexts is the self-serving bias, which refers to the tendency to overestimate our contributions to the team efforts while underestimating others' contributions. For example, suppose that husbands and wives are asked to estimate their proportional contribution to household chores (cleaning, fixing, mowing, doing dishes, taking out the garbage, etc.). Typically, both claim to be doing more than 50 percent of the work (a mathematical impossibility!) (Ross & Sicoly, 1979). Similarly, if two siblings are asked how much responsibility they take in caring for elderly parents, often both will claim to be doing the majority of the work (Lerner Summers, Reid, Chiriboga, & Tierny, 1991). The outcome of a team's efforts also has an important effect on how we evaluate our contribution to it (Miller & Ross, 1975). If the team succeeds, we are quick to assume that our contribution was integral to its success. If the team fails, we attribute that to the undue influence of others.

Thus we see in humans that the essential ingredients for effective teamwork coexist with the seeds for undermining its full potential. Destroying those seeds before they sprout was the moral quest of a particular order of Catholic monks, the Irish Christian Brothers, and rugby was the hammer they used to accomplish their mission.

The Old Christians

Inculcating a strongly communal, rather than individual, sense of self was central to the Irish Christian Brothers' approach to building good Catholic character. In the early 1950s, the brothers were approached by a group of parents from the Carrasco suburb of Montevideo, Uruguay, with the idea of starting a boy's school. Most of the parents were conservative Catholics who were alarmed by the increasingly secular, atheistic bent of the public schools and who were wary of the Jesuit schools with their emphases on intellectualism and social activism. The Christian Brothers were more to their taste, as they stressed unblemished Catholic doctrine and character development. The brothers accepted the invitation, and in 1955 they arrived in Montevideo and opened Stella Maris College.

What the Carrasco parents didn't know was that to the Christian Brothers, building character meant learning to play rugby. Dismayed that their boys were being taught this strange, foreign sport and not the Latin American passion, soccer, the parents protested. But on this point, the monks would not bend. Soccer, they declared, was too individualistic. It taught the wrong lessons about manhood, about virtue, and about being a good Catholic. For that, only rugby would suffice. Eventually, not only did the parents acquiesce, but they and their sons became rugby enthusiasts. The boys enjoyed the game so much that after they left the school, they wanted to keep playing. In 1965, Stella Maris alumni formed the "Old Christians" rugby club, dedicated to playing rugby on Sunday afternoons, and in 1968 and 1970, the Old Christians were Uruguay's national champions. In 1971, the Old Christians traveled to Santiago to play the Chilean national team. The trip was such a success that plans were made for a rematch the next year. And so it was that in October 1972, the team, their families, friends, and supporters boarded UAF flight 571 in Montevideo to go to Chile for a much anticipated holiday of rugby and touring Santiago. They never got there. Their match would be against a far different opponent.

CHAPTER TWO

The Evolution of Taboo

SCENE 2: STARVATION

As Nando Parrado recounts in *Miracle in the Andes,* less than a week into their ordeal, he found himself standing outside the wrecked fuselage of the plane, staring at a single chocolate-covered peanut, his last morsel of food. He put the candy in his mouth but allowed himself only to gently suck off the chocolate, saving the peanut for tomorrow. The next day, he split the peanut, nibbling ever so slowly on just one half, making it last for hours. On the third day, he did the same with the other half. When that was gone, a grim reality took hold—there was no food left. Lying in the darkness of the battered Fairchild, Nando whispered cautiously to Carlitos Paez about using the bodies of the dead for food. "God help us," Carlitos whispered in reply, "I have been thinking the very same thing." (96)

Almost immediately after the crash, team captain Marcelo Perez took control. He quickly organized efforts to free the trapped passengers. He set the strongest boys to work clearing and shoring up the plane's interior, their only protection against the approaching bitter cold of night. Over the next few days, Marcelo formed teams with different responsibilities: a medical crew composed of Roberto Canessa, Gustavo Zerbino (two medical students), and Liliana Methol, their self-appointed nurse; a cabin cleanup crew made up of many of the younger boys; and a water-making crew composed of the injured and those too weak for anything more strenuous. Marcelo's early leadership was lifesaving. Slowly, under his direction, the traumatized group pieced together some order from the chaos.

For himself, Marcelo reserved inventorying and rationing their resources. Since most of the supplies and luggage had been stored in the now-detached tail section, what remained was meager: three bottles of wine, a bottle of whiskey and another of cherry brandy; eight bars of chocolate; two cans of mussels and one can of almonds; some crackers, dates, and dried plums; and a few small jars of jam. These paltry items he parceled out to the twenty-eight survivors. Lunch was a deodorant cap of wine and a taste of jam, and dinner was a square of chocolate. Slightly larger rations were allowed on Sundays.

Hunger hit them hard and fast. The conversation in scene 2 took place only four days after the crash. The high altitude compounded their physical deterioration. Their early expeditions through the hip-high snow became slow-motion duels with death; escaping the mountain seemed impossible. By the tenth day, their pieced-together radio told them that rescue efforts had been halted. No one would save them. Abandonment and starvation forced them to openly confront what heretofore had been the exclusive purview of private thoughts or cautious whispers—eating the dead.

As a second-year medical student, nineteen-year-old Roberto Canessa assumed the role of the group's chief medical authority (Zerbino was a first-year student). When the terrible necessity of using the dead as food could no longer be ignored, Canessa took the lead. Using his medical knowledge, he passionately pleaded the clinical reality of their plight: in the absence of adequate nutrition, their bodies were steadily self-consuming in order to continue functioning. But his argument was not just medical; it was theological as well. God had given them the means of survival; not to use it would be committing the sin of suicide. Although this made sense, it did nothing to diminish the grotesque nature of what they were contemplating. It was left to Canessa to smash the taboo so that others could follow:

> Canessa took it upon himself to prove his resolution. He prayed to God to help him . . . [but] the horror of the act paralyzed him. His hand would neither rise to his mouth nor fall to his side while the revulsion which possessed him struggled with his stubborn will. The will prevailed. The hand rose and pushed the meat into his mouth. . . . He . . . had overcome a *primitive, irrational taboo.* He was going to survive. (*Alive,* 81–82, italics added)

Taboo is primitive if we understand "primitive" to mean "primary, early emerging, ancient in origin." Taboo has a deep evolutionary history, which gives it a powerful hold on our emotions. But it is not irrational if we understand "irrational" to mean "without reasons." Taboo has reasons—good reasons, but so old that sometimes they seem badly out of touch with current circumstances.

Our minds are intimately interconnected with our emotions; in fact, emotion is part of the mind. Our emotions have evolved as a guidance mechanism for our minds. Our minds "trust" our emotions to illuminate right actions. For example, that queasy discomfort we experience when we agree to lie for a "friend" serves as a nagging warning about the potential moral consequences. The intuitive and often emotion-laden thinking that produces reflexive moral judgments has been called the mind's "system 1" (Cummins & Cummins, 2012; Kahneman, 2011). System 1 processes are often based on experience and well-honed problem-solving heuristics and thus can be very effective for keeping us in good social standing. Taboo is powerful because it activates the system 1 processes that warn of moral and social danger.

Sometimes, however, our immediate emotional reaction may not be entirely morally trustworthy, prompting us to engage the more systematic, deliberative processes of system 2. When these two systems clash—when our gut tells us one thing and our calculations tell us another—we become deeply conflicted. What we *conclude* is right nonetheless *feels* very wrong. Accordingly, to break this taboo, Canessa had to reject the dire warning of system 1—that this was a disgusting act of disrespect for the dead that would make him a social outcast. Instead he had to rely on the calculations of system 2—that this was the only way to survive, that it was morally right to survive, and that the cultural restrictions of the civilized world did not hold here (Cummins & Cummins, 2012).

Taboo

The term "taboo" is derived from the Polynesian word "tabu," meaning "forbidden" or "to forbid" (Radcliff-Brown, 1952). English voyagers to the Polynesian islands adopted this term to describe the many seemingly strange prohibitions they encountered among the native people. These prohibitions included both actions (saying certain names, eating certain foods) and things (certain places or persons). For instance, if someone touched a corpse, he would become taboo. This meant that he had entered a precariously vulnerable state, and unless he scrupulously regulated his behaviors in seemingly arbitrary ways, such as not eating with his hands, he would likely fall ill and maybe die (Radcliff-Brown, 1952, 133–134). Often taboo deals with entities and situations that a given society considers odd or anomalous and therefore potentially threatening. For example, anthropologist Mary Douglas (1966/1978) argued that pigs were taboo to the ancient Hebrews because they failed to meet their criteria for a "normal" farm animal (it must have cloven hooves and chew its cud).

To outsiders, the "oddness" of taboo prohibitions link them with irrational, superstitious practices. Indeed, it was originally thought that

"rational" Western societies were free of taboos and that they were restricted to the "brown and black races of the Pacific" (Radcliff-Brown,
1952, 134). Every society, however, has its prohibitions that appear odd
or superstitious to outsiders: Catholics are forbidden to eat meat on
Fridays during Lent.

None of us is entirely immune to the emotional pull of "irrational" prohibitions. Suppose you found out that your favorite, fur-lined,
black leather coat was once worn by Adolf Hitler. Would you continue
wearing it? Psychologists Carol Nemeroff and Paul Rozin found that
people recoiled in disgust at the idea of wearing clothes once donned
by a mass murderer, even after those clothes had been thoroughly
laundered (Nemeroff & Rozin, 1994). Likewise, 80 percent of college
students interviewed believed that simply wearing one of Mr. Roger's
sweaters transmits to the wearer a bit of the TV star's humane essence
(see *Psychology Today* March 1, 2008). This, they thought, would happen even if the wearer were unaware of the sweater's past history.

These findings highlight a very natural, pervasive, and adaptive trait
of human thinking, something called "magical contagion," the notion
that by their association with someone or something, objects or artifacts can possess and transmit some "essence" of that person or thing.
This essence is often evil or dangerous, but not always. Magical contagion is also what attracts us to holy relics or persons. Despite our
level of education or sophistication, we never fully extract ourselves
from the primal emotional grip of magical contagion. Take a piece
of chocolate and form it into the shape of a turd—still want to eat it?

In our ancestral past, people did not understand "germ theory" or
the science behind infectious disease, but they did understand invis-
ible transmission. They noted that touching a sick, dying, or dead person might precede illness, so they often developed ritualized prohibitions to avoid such hazards. We observe the same prohibitions today,
only under a different set of assumptions. You should not touch (it is
taboo to touch) a dirty floor or counter before handling food. If you

do (and therefore become invisibly contaminated), you should apply a special lotion to your hands (soap) and repetitively (ritualistically?) rub them under running warm water in order to restore your purity.

Taboo, however, involves more than just practical avoidance strategies. It is practical avoidance cloaked in sacredness. Human nature is such that compliance with practical restrictions is always elevated if a powerful authority is believed to be behind them. Although simply telling restaurant employees to wash their hands helps, punishing violators with lost pay or termination works even better. Philosophers from Aristotle to Marx observed that moral obligations are always more readily followed if divine command, rather than just human reasoning, lies behind them. This is probably why anthropologists have found that sacredness—the idea that some objects, rules, ideas, traditions, and the like are backed by a transcendent authority—is a universal belief of human societies (Radcliff-Brown, 1952). To transgress sacredness is to summon other-worldly wrath. Taboo, then, takes what might have started as a practical danger-avoidance strategy and puts divine authority behind it. Its violation takes on a whole new dimension. No longer is it just taking a personal risk; it is flouting the very order on which the society is based.

Taboo and the Making of Human Society

Taboo concerns what individuals may and may not do. In today's world, taboo is seen as irrationally and arbitrarily restricting individual freedom and constraining self-expression and personal gratification. If the popular media are to be believed, taboo is most useful in being broken, not in being observed. The universality of taboo suggests that historically at least, taboo performed some (now largely outdated) important social function. This is true. I shall try to convince you that taboos performed (and may still perform) a crucial social function; that is, they created the conditions for the very existence of human society.

All humans have values, which, in this instance, refer to acting in ways that specify interests (Radcliff-Brown, 1952, 139–140). For example, suppose that I spend thousands of dollars on season football tickets; regularly visit the ESPN website to check scores, scour statistics, and read articles on football; and arrange and rearrange my schedule so that I can attend games or watch them on television. From this, you could reasonably surmise that I value football. Football represents a personal value of mine, something that may or may not be shared by others. Herein lies the important distinction between a personal value and a social value. A social value must be shared by more than one individual. If I start seeing a special gal, she'd better share my interest in football; otherwise our relationship may be in jeopardy. Stable long-term relationships must be predicated on shared values. If my gal and I agree that music is life enhancing, family comes before money, and Sunday football is sacred, then we might share enough values to make a good match. Among intimates and close kin, shared values can usually be signaled directly through talk or observation. But when human groups expand beyond just close kin, how are shared values established, reinforced, and signaled so that we all know which ones our group holds dear? Taboos can help.

Signaling Group Values

The hill country stretching north to south through what is today modern Israel is home to scores of Bronze and Iron Age archaeological sites of ancient kingdoms like Canaan, Judah, Israel, Moab, and Samaria. Archaeologists search these hills for the origins of the ancient Israelites (Finkelstein & Silberman, 2001). The myriad different tribes who once trod this weathered ground—the Edomites, Hivites, Moabites, Ammonites, Jebusites, Amorites—are now just arcane memories. Archaeological sites preserve the remains of these and countless other forgotten peoples. Why the descendants of the ancient

Israelites have remained a viable, defined community while so many of their historic contemporaries have faded away is a fascinating and complex issue that cannot be reduced to a single, simple factor. Nonetheless, one undoubtedly relevant factor is that the children of Israel have stubbornly retained a set of cultural markers distinguishing them as a community.

Orthodox Jews are easy to pick out of a crowd. They wear long dark coats and hats; have conspicuously long, braided hair; and often pray publicly, rhythmically bowing and swaying. Their lifestyle is filled with prohibitions—no work on Shabbat, no non-kosher food, no touching menstruating women.

Among the First Iron Age (roughly 1150 to 900 BCE) archaeological sites in the hills bordering what was ancient Canaan, a few stand out for what is conspicuously absent from their refuse—pig bones. Earlier sites contain pig bones, and contemporaneous sites from coastal regions also have a plethora of pig remains. So why are there no pig bones in just these particular sites? Archaeologists Israel Finkelstein and Neil Silberman (2001, 119) argue that these sites could point to the origin of the ancient Israelites, a people set apart by a peculiar taboo: no eating pork. This and other prohibitions (e.g., no worshiping foreign gods) did more than just arbitrarily mark certain people as different from others; they represented the group's shared values.

Jewish dietary restrictions concern food only secondarily. They are primarily about respecting Jewish law—Torah (Donin, 1991). Not doing what every else does routinely—eating pork chops, marrying whomever you want, dressing in the latest fashion—signals to others that you are serious about the Torah. If others wish to form a serious, long-term relationship with you, then they had better share that value as well. Torah is a shared value strong enough to bind communities, and its taboos are public "badges" that people wear to signal their commitment to its values.

Jews are hardly alone in this. The Andaman Islands form a north-south archipelago stretching across the Bay of Bengal between India and Burma (Myanmar). Recent genetic studies indicate that the Andaman Islanders may be descendants of some of the original "out of Africa" migrants dating back 60,000 YBP (Endicott et al., 2003). Among the many taboos of Andaman society are those dealing with names that must not be spoken—those of pregnant women, the recently dead, those mourning the dead, those recently married, and youths going through adult initiation (Radcliff-Brown, 1952, 146–148). The reason for this appears to be similar to that discussed earlier for Polynesian taboos, that is, the vulnerability of those in precarious or transitional states. Note that all those included in the Andaman name taboos are in some transitional status—expectant mothers, the recently dead, the recently married, those in mourning, and adolescents becoming adults. Names hold a special power in Andaman society (as is true in many traditional societies). They don't simply identify persons, they capture their essence, personality, and even their very existence as social beings. But while a person is in transition, all these qualities become unstable, which makes the person highly vulnerable. Thus, using the name could endanger him or her, so by observing the taboo, others express their understanding of and concern about the person's vulnerable status. One person's transitional vulnerability thus becomes a shared concern or shared value for the entire community.

Something similar can be observed among the Lohorung Rai, a traditional agricultural people living the eastern highlands of Nepal. Immediately after birth, the mother and newborn are considered to be in a vulnerable transitional state, during which they are especially prone to arousing supernatural anger (Hardman, 2000). Accordingly, they are subject to a host of taboos concerning what they can and cannot eat or touch and who can and cannot touch them. Mother and baby remain secluded for five days until the baby can be ritually introduced to the ancestors. During this seclusion period, it is also taboo for anyone in the entire Lohorung Rai village to conduct ancestral rites. Thus the entire community shares in the fate of the mother and her newborn. They become a shared value of everyone.

These examples highlight two important social functions of taboos. First, taboos symbolize the shared values of a community, thereby allowing community members to participate in the expression of those values (Weber, 1963). This both defines and unites a community. Observing taboos becomes a public signal saying, "This is important to us," "This is what we share." Taboos take individuals beyond their narrow personal values and identify for them the broader set of common values that makes them part of a community. In this way, taboos set boundaries: "If you share these values too, then maybe you can be part of us. If not, then you are a stranger."

Second, taboos define what is threatening or dangerous to a community and provide the procedures for avoiding the danger. Avoidance, rather than confrontation, is taboos' great strength. In other words, taboos deal with danger by keeping it at bay rather than trying to overcome it once it has arrived (Steiner, 1956/2004). In our evolutionary past, those groups with effective danger-avoidance strategies very likely had an advantage over those who relied more on confronting and defeating dangers.

Note that one of the common themes of taboos is the precariousness of those in transition. Transitions threaten the social order. Nothing is more disruptive than an upheaval in the established order, such as when a new male challenges the alpha male for supremacy or when a subordinate matriline tries to topple a more dominant one. Social transitions can be violent and dangerous and can put the group's very existence in peril, putting all group members at risk. Taboos (along with related rituals and ceremonies) specify a set of procedures for regulating social transitions. Births, deaths, marriages, and maturation are social transitions in whose outcome the entire community has a stake (Van Gennep, 1960). Avoiding the danger that these transitions could produce is highly preferable to extricating the community from the danger once it has emerged. By following taboos and other cultural traditions, community members join with and demonstrate support for those in transition while at the same time those in transition demonstrate their continued commitment to the group and its values. These public demonstrations of commitment enhance trust among the community's members (Sosis, 2004).

Breaking Taboo

What happens when a taboo is broken? If upholding a taboo builds trust and community, then breaking it is tantamount to betrayal. Parrado describes how, in their starved condition, he and his friends were ashamed to be caught staring at the bloody wound of one of the boys (*Miracle in the Andes*, 95). Their embarrassment was not because of anything they had done; instead, it was because of what they were thinking. Though unwanted and involuntary, the thoughts were supremely dangerous. They were the unthinkable thoughts that preceded the violation of a taboo.

When does a friend stop being a person and become a piece of meat? If we start down that path, where will it end? If we eat the dead, will we then contemplate eating the nearly dead? What about the injured? If death is inevitable for some, is it wrong to hasten the process in order to fend off starvation? Will the entire social fabric unravel if necessity requires us to pull out a few threads? Even the divine law bolstering a taboo may not forestall the calamitous momentum. Divine laws, after all, are rarely absolute. "Thou shalt not kill" does not apply to enemies in war, but does it apply to friends when we are starving? The guilt and embarrassment associated with breaking a taboo arises from the violation of trust inherent in the act. Friends don't stab one another in the back (literally) for a meal, regardless of how hungry they are. Just thinking about crossing that line was a precarious step toward every man for himself, an abandonment of their shared humanity.

To protect ourselves from the most disastrous consequences of violating taboos, humans have evolved a taboo mentality; that is, we are highly suspicious of those who even contemplate breaking the taboos representing a group's sacred values. As an example, consider the following: One day, you jokingly ask your mother if she ever thought about selling you into slavery when you were a little child. She pauses for a moment and then smiles and says, "Of course not, dear." Are you a bit shaken? Do you start to question your mother's love? The only correct answer to such an inquiry (and the answer you expected and wanted) was for her to take immediate offense—"My god, dear, how could you even ask something like that?"

Philip Tetlock, a psychology professor at the University of California at Berkeley, does research on the taboo mentality. In one study, he presented people with scenarios in which a hospital administrator had to make difficult decisions (Tetlock, Kristel, Elson, Green, & Lerner, 2000). One scenario featured what was called a tragic trade-off, in which neither choice was good but a choice had to be made. Thus, because of limited resources, the administrator was forced to choose between saving the life of one of two boys. Most people judged the administrator more positively when he took a long time, rather than a short time, making his decision, regardless of which boy he decided to save.

In a second scenario, the administrator had to make a taboo trade-off. That is, he had to decide whether to save the hospital a million dollars or to save a boy's life. This was a taboo trade-off because it pitted something sacred (a human life) against something mundane and secular (money). Would people regard it as taboo to even contemplate putting human life on the same level as money? Apparently yes. Merely thinking about the decision for an extended period of time caused people to judge the hospital administrator negatively even if he ultimately decided to save the boy. The lack of an immediate, emotional revulsion to equating the sacred and the secular made him suspect.

By contrast, the administrator was judged most positively when he decided quickly in favor of the boy. In this instance, reflexively upholding the taboo of "don't even consider human life and money as comparable" was the administrator's best strategy to maintain his reputation. A powerful message of these results is that cementing trust in others is less a matter of assessing how they act than of making sure that they have the right emotional barriers preventing them from even entertaining certain objectionable thoughts. In other words, do they have in place the right emotional guidance system? Taboo is about avoiding danger, and the danger of being considered untrustworthy is best avoided by vigorously adhering to the taboo.

Once a taboo has been violated, even inadvertently, atonement must be sought. We must reassure both others and ourselves that our action was an anomaly, not indicative of an enduring disposition. For example, suppose you consider yourself a liberal egalitarian in social and political matters and are faced with information showing a correlation between race and the incidence of fires in a neighborhood (e.g., fires occur more often in a minority-race section of town). More than likely, despite the correlation, you would probably still object to using a racial criterion to determine the rates for fire insurance (so-called redlining).

In their study, Tetlock and his colleagues (Tetlock et al., 2000) had liberal egalitarians role-play insurance executives who consented to redlining. Once they recognized their violation of a taboo, the liberal egalitarians sought to "absolve" themselves by volunteering for anti-racist causes. This was true even when they violated the taboo inadvertently. Furthermore, the researchers found absolution-seeking behavior even in those who had only witnessed, but failed to condemn, the violation. What these studies show is that simply being a party to the violation of a taboo, as either an unintentional perpetrator or a silent witness, is enough to produce an aversive emotional reaction. That reaction, most likely a feeling of guilt or shame, serves as a guidance mechanism directing us to engage in corrective actions to repair both our self-image and our public reputation.

Evolutionary Origins of Taboo

The original "Polynesian" standard for taboo was an automatic, un-reasoned, and absolute aversion to breaching a perceived sacred barrier (Tetlock, 2003). Using that criterion, it is hard to find anything in the nonhuman world that can truly be called taboo. Animals simply don't have "sacred" barriers that "absolutely" should not be crossed. Alpha males should not be challenged, and most of time they aren't, but all of them will be challenged eventually. Predators like leopards and pythons should be avoided, and most of the time they are, but not always. Gombe videographer Bill Wallauer documented an intriguing instance in which a group of chimpanzees lingered for tens of minutes in fright and fascination over a python they encountered (see Rossano, 2010b, 129–130). They alternately approached, pulled back, hugged, and reassured one another while also emitting alarm cries. The dangerous and the forbidden always have their allure.

Only with humans can we talk about taboo as something forbidden by virtue of its sacred status. So how might such a notion have arisen over the course of our evolutionary history? We will never, of course, know exactly what the first taboo was. How to hunt, what to gather, what to eat, how to prepare food, and where to sleep—any of these essential activities could have been the origin of the first taboo. Given their direct relevance to fitness, however, sex and reproduction probably were the site of humankind's earliest taboos.

Incest taboos, for example, are present in nearly all human cultures, and all our primate relatives have inbreeding avoidance mechanisms such as male or female dispersal from natal groups. However, having sex with the "wrong" partner has direct negative consequences for individual fitness and therefore is relatively ineffective as a signal of commitment to group values. In other words, people could follow incest taboos for purely selfish reasons. Accordingly, a more probable circumstance for the origin of taboo as a signal of shared values is what comes after sex: birth.

Arguably, there is no activity more directly related to fitness than birth: the very act of reproduction itself. No primate has a more challenging and potentially life-threatening birthing process than humans. Thus, there is no other activity for which taboos carry higher evolutionary stakes. In traditional societies, the taboos surrounding birth are nearly universal, and many of them seem quite arbitrary. There also is evidence indicating that the perils of birth emerged very early in our ancestral history.

Footprints in Dust

About 3.5 million years ago, a few of our ancestors walked across a patch of land in north-central Tanzania at a site that today is referred to as Laetoli. Volcanic ash preserved their footprints, which were discovered in 1978 by a team led by Mary Leakey. The footprints were made by hominins known as *Australopethicus afarensis*, a species that arose about four million years ago, only about two million or three million years after the origin of the hominin evolutionary branch. The footprints show unequivocally that *A. afarensis* was bipedal—that it walked upright on two legs and not on all fours. An analysis of the footprints confirms the unambiguous humanlike nature of *A. afarensis*'s gait (Crompton et al., 2011). It shows that when *A. afarensis* walked, it did so in a way more similar to how humans walk than how chimpanzees or other apes walk when they assume a bipedal stance. Walking upright was natural to *A. afarensis*, not an odd posture occasionally assumed by an otherwise quadrapedal ape. This analysis has been reinforced by another one, which looked at *A. afarensis*'s foot bones (Ward, Kimbel, & Johanson, 2011). The bone structure is similar to that of humans, supporting the notion that *A. afarensis*'s gait and stance, while not exactly human, was fairly similar.

Why is humanlike bipedalism relevant to taboo? Fully committed bipedalism, like as that found in *A. afarensis*, constrains pelvic dimensions (Rosenberg & Trevathan, 2002). These constraints can mean a more difficult birth process. How much more difficult depends largely on the size of the infant that is trying to pass through the birth canal.

One study indicates that at around this same time (3.5 MBYP), newborn babies' head and body size increased significantly (DeSilva, 2010). On average, human mothers give birth to infants that are about 6 percent of their body mass. By contrast, chimpanzees give birth to infants that are only about 3 percent of their body mass. An analysis of the fossil remains of a range of different hominin species showed that between *Ardipithecus ramidus* and *Australopithecus afarensis*, infants' body mass jumped. In other words, *A. ramidus* gave birth to infants that were apelike in their infant/mother body mass ratios (less than 3 percent), and *A. afarensis* gave birth to infants that were more humanlike (around 5 percent). This indicates that the complications inherent in human birthing may have arrived quite early in hominin evolution. Indeed, evidence suggests that the size of *A. afarensis* newborns was at or near their mothers' pelvic-opening capacity (Tague & Lovejoy, 1986).

Another complicating factor in human birth is head rotation. For humans, the newborn's head must rotate as it passes through the birth canal so that it ends up facing away from mother as the baby is born. This, however, makes it hard for mother to guide the baby out, making assistance even more necessary. Humanlike head rotation was probably not part of the *A. afarensis*'s birth process. Currently, the earliest potential evidence for head rotation is dated to about 600,000 YBP in a later *Homo* species known as *Homo heidelbergensis* (see the discussion in Franciscus, 2009). DeSilva's analysis thus may overestimate when hominin birth became similar to human birth. But there is no question that this occurred at some point in human evolution and that our ancestors had to find a way to deal with it effectively. Not surprisingly, then, in traditional societies, birth has many taboos.

Traditional Birth

Our understanding of human prehistory is inevitably cloudy and incomplete. Although approximate and imperfect, extant traditional societies give us a glimpse of that world. Finding commonalities across these societies hints at what might have occurred in our past. One commonality is the pervasiveness of rituals and taboos associated with pregnancy and birth. Food taboos are nearly ubiquitous (Stanton, 1979).

For example, among the Orang Asli (a generic term for a number of aboriginal tribes in west Malaysia), pregnant women are restricted to eating small animals believed to have weak spirits, like rats, toads, small birds, and fish (Bolton, 1972). Furthermore, fish and rodents may be eaten only if caught by the pregnant woman's husband or a close relative. After giving birth, the mother is restricted to a gruel diet for a week. Among the Trobriand Islanders of Kiriwina, pregnant women may not eat fish that attach themselves to coral, nor may they eat fruits such as bananas or mangoes, as all are thought to cause difficult birth or birth defects (Malinowski, 1922; 1929). The Onabasulu of Papua New Guinea ban eggs from a pregnant woman's diet (Schieffelin, 1976), while in traditional Samoan society forbids pork legs or any other food considered appropriate for the senior males of a household (Barclay, Aiavao, Fenwick, & Papua, 2005). Finally, among the Lohorung Rai, a pregnant woman or new mother must eat certain "hot" foods such as chicken, meat, honey, or rice, and she must drink "hot" drinks like chicken broth or warmed millet beer (Hardman, 2000). Other "hot" foods, such as pork, buffalo meat, chilies, and other spicy foods, however, are prohibited. "Hot" and "cold" in this context refer less to the food's temperature and more to its symbolic power. A person's health depends on maintaining a balance between hot and cold, and too many cold foods are thought to make the baby retreat into the womb.

The birth process itself also is subject to numerous prescriptions and proscriptions. Among the XKo—"bushmen" of Botswana, southern Africa—before a woman gives birth, her hands and body, excluding her genitalia, must be washed by her attendants (Heinz & Lee, 1979, 140–141). One of the attendants kneels facing the mother. Her job is to "catch" the baby, and her hands must be washed in the juice of special roots. Once the baby is born, the umbilical cord must be cut with a special instrument, a sharpened grewia stick hardened in fire. It must never be cut with a knife. The cord must then be buried deep enough so that hyenas or other creatures cannot get it; otherwise the baby will die. The cord's burial place is marked with thorns and branches. The afterbirth, blood, and placenta also are buried in this hole.

The cultural ideal among the !Kung San of southern Africa is for a woman to give birth alone in the bush (Thomas, 1989, 156–159), although a woman's first birth or a difficult birth often involves attendants. Ideally, when labor begins, a !Kung woman simply slips into the bush without telling anyone. It is believed that expressions of fear or pain indicate that the woman really does not want the baby, and so a !Kung woman is not to cry or shout out in agony during labor but instead may only clench her teeth or bite her hand, even to the point of bleeding. In anticipation of the birth, a !Kung woman prepares a small mound of soft grass where the baby is to be born. Once the baby is born, the mother is to saw off the cord using sticks and then she must wipe the cord clean with grass. All the blood-stained grass, along with the placenta, must be covered (not buried) with stones and branches and clearly marked so that no man will step on or over the spot. The spot is believed to be tainted with a power that could cause a man to lose his masculinity and hunting ability. Burying the placenta is forbidden, as this could cause the mother to become barren.

The !Kung's emphasis on solitary birth is not typical, however, as in most traditional societies, assisted birth is the norm (Stanton, 1979). Often, tradition dictates who the assistants can and cannot be. For example, the XKo's birth attendants are usually the expectant mother's mother, her mother-in-law, and the woman who instructed her during her adult initiation. Males are excluded. The birth attendants in many traditional societies are older female relatives, although the exclusion of men varies. Some traditional people like the !Kung, Lohorung Rai, and Ebrie (Ivory Coast, West Africa) exclude males. But others, such as the Ache (Latin America), permit husbands to be present at the birth (Konner, 2010, 400–401).

The universality of traditional societies' birthing taboos, coupled with the early emergence of birthing complications in hominin evolutionary history, suggest that birth may have been one of the first places where taboos arose. These taboos probably helped ensure that certain practices were followed, practices thought most likely to result in a successful birth.

Taboos involve more than just prohibitions or prescriptions; they also must be linked to the supernatural. Note that the supernatural still is present in many of the present-day hunter-gatherers' birthing rituals (e.g., if the umbilical cord is not buried deeply enough, then the baby will die; if a man steps on the covered placenta, then he will no longer be able to hunt).

Can we make any reasonable speculations about the origins of the supernatural aspect of taboos? I believe that we can, and my argument is based on four observations. (1) The origin of supernatural thinking lies (at least in part) in childhood imagination; (2) "neoteny" (the process of incorporating juvenile traits into an organism's mature form) has played an important role in human evolution; (3) incorporating such aspects of childlike thinking as playfulness, imagination, and curiosity into adult cognition can be socially advantageous; and (4) the adult social world had become increasingly complex by sometime between 300,000 and 70,000 YBP, and this provides the most likely time when supernatural thinking would have been incorporated into hominins' social functioning.

SUPERNATURAL THINKING AND CHILDHOOD IMAGINATION

Studies in developmental psychology clearly show that nearly all the imaginative features associated with supernatural religious beliefs can be found in childhood cognition. Children readily accept or create notions about omniscient beings with suprahuman powers (Barrett, Newman, & Richart, 2003; Barrett, Richart, & Driesenga, 2001). They view magic as a legitimate causal force operating alongside other natural physical causes (Harris, 2000, 162–166; Phelps & Wooley, 1994). They see design, purpose, and justice as inherent in the natural world (Fein, 1976; Jose, 1990; Kelemen, 1999a), and they insist that some mental functions persist beyond death (Bering & Bjorklund, 2004). So pervasive are these qualities of childhood thinking that some researchers believe that children are "intuitive theists," naturally prepared to accept supernatural concepts (Barrett & Richart, 2003; Kelemen, 2004). One reasonable conclusion therefore is that the supernaturalism that supports taboos may very well have had its origins in childhood imagination.

NEOTENY

Human evolution is marked by *neoteny*, the persistence of childhood traits into adulthood, which makes the adult form of a descendant species similar to the ancestral species' infant or juvenile form (Gould, 1977; Konner, 2010, 61–62; Parker & McKinney, 1999). For example, human adults have flattened faces (the lower face does not protrude outward, as adult nonhuman apes' faces do) and hairless bodies. Both of these are true of infant apes but not of adult apes. In this instance, the infant ape is taken to be representative of what our juvenile ancestors (e.g., *A. afarensis or H. erectus*) would have been like. Thus, an important developmental change that occurred over the course of human evolution was that we retained certain juvenile traits in the adult form, rather than allowing those traits to "mature," as they did in our ancestors.

The role of neoteny in human evolution has been a matter of debate. For some researchers, such as Dutch anatomist Louis Bolk, neoteny was so pervasive that humans could be considered a sexually mature fetal ape (Gould, 1977, 361). Bolk's enthusiasm for neoteny was echoed many years later by the well-known paleontologist Stephen Jay Gould in his book *Otogeny and Phylogeny*. Others have argued that the importance of neoteny is overplayed and that many of the most important traits distinguishing humans, such as our large brains, actually resulted from a process of overdevelopment (McKinney & McNamara, 1991; Parker & McKinney, 1999). By overdevelopment, these researchers mean that the growth rate of the human fetal brain was extended further into childhood than it is in other primates, thus producing a larger brain. In the end, the adult human is a mosaic of juvenilized and overdeveloped features (Konner, 2010, 139), and among the juvenilized traits are mental ones pertaining to social skills.

CHILDLIKE THINKING AND SOCIAL COGNITION

Retaining some juvenilized mental traits into adulthood could very well have served an adaptive purpose, as imagination improves the social skills of both children and adults. Parker and McKinney (1999, 290–291) contend that pretend play may have evolved in human children as an adaptive strategy for acquiring technical skills (hunting, cooking) and social roles (mothers, fathers, hunters, ritual leaders, etc.). This is consistent with anthropological studies showing that among traditional societies, play is central to skill development and enculturation (Konner, 2010, 637–638). Typically, boys and girls in hunter-gatherer societies learn important skills and acquire their social identities through observation, imitation, and play, with only a modicum of active instruction.

There also is empirical evidence showing that more imaginative children have better social and cognitive abilities. Children who more frequently engage in role-playing score significantly higher on tests measuring the understanding of others' mental states and how others' thoughts, emotions, desires, and beliefs vary based on situational factors. Preschoolers who engage in more pretend-play role-playing are seen as more likable and receive higher sociability ratings from peers and teachers. Finally, increased engagement in simple make-believe activities, either alone or with others, has been positively linked to understanding others' mental states (Connolly & Doyle, 1984; Howes, 1988; Lalond & Chandler, 1995; for a review, see Harris, 2000, 30–31). Enhancements of social intelligence have also been linked to another category of pretend play: imaginary companions.

Cross-culturally, as many as 65 percent of seven-year-olds have or have had imaginary friends or pretend-play partners (e.g., stuffed animals or toys endowed with personality). Children with imaginary companions are often precocious on a number of measures of social awareness and empathy. For example, these children are typically less shy, more sociable, and more emotionally expressive, and they tend to score higher on theory-of-mind measures than do children without imaginary companions (Taylor, 1999; Taylor, Carlson, Maring, Gerow, & Charley, 2004). In addition, children with imaginary companions are able to produce more complex sentence structures and are more capable of referential communication than are children without imaginary companions. "Referential communication" refers both to the ability to recognize what verbal information another person needs and to express that information effectively (Bouldin, Bravin, & Pratt, 2002; Roby & Kidd, 2008).

Whereas social/cognitive enhancements are associated with imaginary companions, their absence has been associated with deficits. The lack of imaginary friends has been associated with poorer performance on measures of emotional understanding. Furthermore, children who do not pretend-play, role-play, or have imaginary companions are at higher risk for autism, a serious social/cognitive deficiency (Baron-Cohen et al., 1996).

The social/cognitive benefits of imagination continue past childhood into adolescence and adulthood. Teenagers who refer to imaginary friends in their diaries tend to be more socially competent and to have better coping skills (Seiffge-Krenke 1993, 1997). Adults who frequently read fiction are more empathetic, are better able to read emotions from facial expressions, and rate higher on other measures of social acumen. These findings have been associated with the fact that when people read fiction, they typically become absorbed in the author's narrative world, vicariously experiencing events from the protagonist's perspective. Conversely, those who read a lot of nonfiction are often found lacking in social intelligence (Mar, Oately, Hirsch, dela Paz, & Peterson, 2006; Oately, 1999).

SOCIAL COMPLEXITY AND HUMAN EVOLUTION

A popular theory regarding the evolutionary origins of human intelligence states that the demands of a complex social world selected our ancestors for their large brains and a uniquely powerful set of mental skills, including a theory of mind, episodic memory, symbolism, and language (Alexander, 1989; Dunbar, 1996; Geary 2005). It is in this complex social milieu that retaining a childlike curiosity, imagination, and playfulness would have benefited adults by making them more socially skilled and adaptable. When might this complicated social world have arisen? Based on the relationship between neocortex size and group size, Aiello and Dunbar (1993) contend that sometime between 500,000 and 250,000 years ago, the size of the average hominin social group grew significantly. It may also be relevant that evidence of pigment use emerges in the archaeological record around this same time (about 300,000 YBP, Barham, 2002). One potential use of pigments is as a social marker by particular groups or tribes coloring themselves in distinctive ways. The first evidence of intergroup trade, however, does not appear in the archaeological record until about 70,000 YBP (Ambrose & Lorenz, 1990; Jacobs et al., 2008). About this same time (70,000 YBP), the first evidence of supernatural ritual may also be present (Minkel, 2006).

Our best estimate for the emergence of taboo therefore is probably 300,000 and 70,000 YBP, when the hominin social world was becoming more complex, with significant increases in group sizes and intergroup trade. In this context, retaining a childlike imaginative capacity into adulthood would have enhanced fitness through greater social intelligence. This intelligence could have been used to solidify social cohesion by attaching a supernatural authority to the behavioral prohibitions surrounding birth and other fitness-relevant activities. In doing so, taboos were born.

Breaking Taboos and Restoring Trust

Six days after Parrado and Paez's whispered conversation recounted at the beginning of this chapter, all the survivors met inside the Fairchild's battered hulk. Whispers became open, but tense, conversation. Put forth were all the rational, medical, and theological arguments for justifying their action: it was their only hope of survival; no one was going to rescue them and they needed energy to trek out of the mountains; God had given them the means of survival, and not to use it would be to commit the sin of suicide; they had an obligation to survive for the sake of their families; and so forth. All were credible arguments, but all were inadequate to mend the wounded trust incurred when breaking this taboo.

At one point in the discussion, Gustavo Zerbino pointed outside at the frozen bodies of their dead friends and asked the group, "What would do you think they would have thought?" It is part of human nature to ask permission when we sense that we are violating an assumed trust. "Would they think we were animals for using them as food?" "Would they think we didn't love them, that we saw them as mere pieces of meat and not as our friends?" In vain, they groped for a way of securing the permission of those they were about to consume. If they could be sure of that, then maybe the sacred trust between friends would not be shattered.

Zerbino did not presume to speak for the dead. But he did speak forcefully for the living. "I know that if my dead body could help you to stay alive," he told them, "then I'd certainly want you to use it. In fact, if I do die and you don't eat me, then I'll come back from wherever I am and I'll give you a good kick in the ass" (*Alive*, 79–80). With this, a pact was formed, that anyone who died from this point on granted permission to those remaining to use his body as food. The body was still sacred, but a friend could freely give himself to others. Although the taboo had been broken, trust was maintained through a new understanding of what it meant to be a friend.

CHAPTER THREE

This Cold and Capricious Place

SCENE 3: MARCELO'S DEMISE

Marcelo Perez sometimes gently scolded Nando Parrado for his pre-occupation with parties and girls. Marcelo's demeanor, by contrast, was purposeful and certain. It was that decisiveness, coupled with a patient humility, that made him such an effective leader on the rugby field. Off the field, he was equally as sure-footed—his life securely guided by his Catholic faith. But when rescuers failed to arrive—when the world and even God seemed to turn their backs on them, Marcelo, was crushed. The orderly world in which he so firmly believed crumbled beneath his feet. Wracked with guilt, he fell into despair and lost the will to lead. On the mountain, all the rules were different. Certainty was a liability; rational order was a deadly delusion. (as told in *Miracle in the Andes*, 110–112)

As the discussion of using their dead friends as food moved to its inevitable conclusion, Marcelo Perez was tormented by a disturbing question: Why would God demand that they do such a reprehensible thing? This was a reasonable question for the devoutly religious and thoroughly decent man that Marcelo was. But for most of the others, that question had become superfluous. Understanding why would have to wait; existential questions about their plight were fruitless. The only relevant issue was survival.

Modern Minds Need to Know

About 4:30 a.m. on January 17, 1994, Los Angeles was hit by a strong earthquake, and a widespread power outage darkened nearly the entire city. Nervous calls flooded into the Griffith Planetarium from worried residents wanting to know why the night sky looked so scary. Patient astronomers assured callers that the sky was perfectly normal; those strange spots of light were simply stars (reported in Consolmagno, 2008, 185). It seems that the more "civilized" our world becomes, the more jarring we find the unexpected. Nature grows increasingly foreign, threatening, and in need of explanation. We demand reasons for the slightest perturbation from the familiar.

Modern societies encase themselves in layers of material and social infrastructure: roofs keep out the rain; heating mitigates the chill; insurance spreads the risk; sewage systems limit infection; levees hold back the flood; and street lights dispel the darkness. All the while, bureaucrats, agencies, lawyers, and engineers construct policies and monitor systems so that we all remain safely distant from danger. Our thinking naturally reflects our socially embedded existence. When there is a calamity—when planes crash, cities flood, plagues ravage, and fires rage—we want to know why it happened. Who was at fault? Where did the system break down? Was the oversight inadequate?

In our uncivilized past, the relevance of such questions was limited. The gods get angry every now and then, nature is violent, and people die; that's just the way the world is and we may never fully understand why. This is not to say that our ancestors did not look for reasons behind events. But as a practical matter, coping and surviving necessarily held a higher priority.

The Myth of a Rationally Ordered World

Marcelo and his fellow Old Christians were products of a Judeo-Christian culture. Numerous scholars have pointed out that the ancient Israelites initiated a worldview foundational to Western ideas of reason, government, science, and commerce (Berger, 1969; Bruce, 2002; Finkelstein & Silberman, 2006; Stark 2003, 2006; Webster, 1986; Whitehead, 1925/1967). Central to this worldview was the notion of a rationally ordered world that was intentionally comprehensible to the human mind and in which humans had a divinely mandated purpose for existing. Nowhere is this more clearly on display than in the Genesis creation myth. While Genesis shares themes and elements with creation myths worldwide (see Leeming, 2010), the Genesis story has an overarching motif that is in sharp contrast with the cultural milieu from which it emerged.

Biblical historians trace the origins of the Genesis creation story to the seventh century BCE, during the time of the Babylonian exile (Westermann, 1994). Babylon had conquered the kingdom of Judah, captured its capital Jerusalem, and destroyed its temple. A significant portion of the Israelite population, including much of the educated class, was forcibly exiled to Babylon, where they struggled to maintain their cultural identity. It was most likely during this period of exile and/or immediately afterward that Jewish scribes compiled the essentials of the creation story. In doing so, they undoubtedly drew on the various oral and written traditions of both Jewish and non-Jewish origin, molding them into a potent cultural defense.

Forcibly immersed in a pagan culture, Jewish scribes emphasized how their view of God and the universe differed from those of their captors. The Babylonian creation myth, the Enuma Elish, tells the story of attempted murder gone badly awry. The father god, Apsu, planned to kill the younger, lesser gods, whom he found bothersome. But he was foiled by his wife, Tiamat, put into a deep sleep, and killed. Then other gods convinced Tiamat that she should avenge Apsu's death by killing Marduk, the leader of the lesser gods. Tiamat and Marduk battle. Marduk won and split Tiamat's body in two, using the parts to create the earth and sky. He then killed Tiamat's new husband, the demon Kingu, whose blood dripped on the earth, creating humans. Marduk then declared that humans must work for the gods.

The World According to the Enuma Elish

The Enuma Elish contains a number of themes distasteful to Jewish scholars: creation as an accidental and flawed afterthought, dismemberment as a source of creation, gods as violent and vengeful, and humans and their concerns as trivial. But the Enuma Elish is not unique; creation stories around the world echo these same themes. For example, the Sumerian gods created the universe after a drunken party, and the results reflect this origin. Bumba, the creator god of the African Boshongo, vomited up creation as the result of a stomachache. The Mayan creator failed twice in making humans, first using clay and then wood. In both cases, the results were destroyed or were turned into monkeys.

A common theme in creation myths is the corrupting activity of a trickster or devil who thwarts or perverts the creator's best intentions (see Leeming, 2010, 355–357). For example, the indigenous Ainu people of Japan believe that the creation process is plagued by a trickster who steals the sun and tries to swallow it. In the Okanagan (Native Americans of the northwest) creation myth, Coyote is the trickster (common among Native Americans). The Old One sends Coyote to establish peace and justice among the earth's people, but instead he divides them into squabbling tribes with different languages. Coyote is also the trickster among the Pima of Arizona. Although he tried to help the creator Earth-Maker, he wound up changing the newly created human into a dog. In the creation story of the Maidu (Native

Americans of California), humans are supposed to have eternal life, but Coyote (again the trickster) failed to heed the creator's instructions and mistakenly brought death into the world. Frustrating and meddlesome as he may be, the trickster actually serves an important purpose in creation myths because he helps explain the flaws in both creation and human nature.

Using body parts or fluids as the source material of creation is another common theme (Leeming, 2010, 310–311, 323–326). For example, the Raven trickster in the creation myth of the Chukchee (or Chukchi) of eastern Siberia flies around defecating and urinating in order to form the mountains and rivers. Among the Kodiak (Alutiiq) people of Alaska, the first woman urinates and spits to create oceans, rivers, and ponds. For the Tonga of Polynesia, the sun and moon were created when two gods fight over a child and split it in two. An ancient Celtic myth describes creation as resulting from the dismemberment of a tyrannical father-god by his offspring, with his skull becoming the sky and his blood the sea.

The use of body parts and body fluids as creative materials appears to serve at least three purposes. First, it connects current creation with an earlier, more chaotic stage in the history of the universe. Although the force behind that chaos still is present, it has been divided and controlled so that life and nature can exist. Second, it creates an intimacy between creation and the gods, whose substance permeates creation itself. This animates creation, imbuing it with spirituality. Creation shares the gods' nature, including their unpredictable, irrational tendencies. It can be cajoled or assuaged (through sacrifice) or angered (by breaking taboos), but as with the gods, it cannot be fully understood. Finally, it connects the present world with the childlike drive to create. Children often delight in creating something separate from themselves using bodily fluids or excrement. As serious as the world may be to us, for the gods it is only a whimsical plaything, something the gods thoughtlessly excreted and with which they sometimes carelessly toy.

Other creation myths confront more directly the gods' nonchalant attitude toward their creation. Issues very important to humans—life, death, and the soul—are often treated quite casually by the gods. To the Blackfoot of the American Upper Midwest, the creator decides whether death will be permanent or temporary for humans simply by tossing a buffalo chip into the river to see if it floats. In the Efik creation story (Nigeria), when the creator becomes jealous of his creation, his wife placates him by bringing death to humanity, which he finds gratifying. The Malagasy (Madagascar) creator god also becomes jealous of humans because he fears they love his daughter more than him. Consequently, he lays claim to human souls, which he takes at his pleasure.

The Genesis Difference

Although the Enuma Elish echoes many common themes found in creation stories throughout the world, for Jewish scribes, this view of the universe and humans' role within it was repugnant. For them, creation was the intentional act of a loving God and reflected his divine law, which was perfect order and reason emerging from chaos. Humans were the pinnacle of that creation, formed in God's own image. It is this notion of an orderly world, intentionally designed to be understandable to the human mind, that many scholars argue was fundamental to the emergence of Western science (Stark, 2003; Webster, 1986; Whitehead, 1925/1967).

Creation stories both reflect and inform a culture's worldview. Certainly, over the course of human history, cultures have varied widely in how they understood the world and what their expectations of it were. Moreover, the Genesis creation story is not alone in positing such things as creation ex nihilo, order out of chaos, or creation arising from divine thought. It is fair to say, however, that no other creation account with these elements has had such an influence on human history. Genesis also has the unusual and attractive element of creation's being repeatedly deemed "good." More emphatically than any other creation story, it encourages humans to expect intelligibility from the natural world. We can ask questions and anticipate reasonable answers. Marcelo Perez wasn't wrong to ask such questions, but they had now become badly out of context.

That our world was intended to be orderly and comprehensible was certainly not a commonly held expectation of our ancestors. Life for them was much more dangerous and unpredictable, and their power to control their own destinies was far more tenuous. Their creation myths thus mirrored their experience. Expecting life to make sense is a far more modern preoccupation.

One of the first of many shocks for the survivors of UAF flight 571 was how brutally cold it was on the mountain. Except for Nando Parrado, none of them had ever even seen snow before. Now they were 11,500 feet high on a glacier. The cold wasn't just physical; it was metaphorical as well. It was nature's icy claw suddenly and violently ripping away civilization's protective layers, leaving them in the grip of a raw, indifferent world little known to most of us but familiar to our ancestors. This cold and capricious place was the past. Survival now required thinking, behaving, and organizing as our ancestors had done.

SCENE 4: THE COUSINS TAKE OVER

A cruel irony—dying of thirst with water all around you. You grab a handful of snow and stuff it into your mouth—a slight relief, but you need more, so much more. So you grab another handful and another; but now your fingers are frozen, your lips are numb, your tongue is throbbing in pain, and you're still thirsty. You are slowly desiccating in the middle of a frozen ocean. Unexpectedly, there's a voice from the back. Thinking out loud he mutters something about the aluminum covers on the seats. If they can be fashioned into a basin-like shape, he says, then filled with snow and pointed toward the sun . . . then maybe. . . . You notice two others—his cousins— have already set to work on the idea. You follow their lead, as do many others. You pause just a moment to sneak a glance at the quiet one in the back who just saved your life. You wonder how many other ideas he might have—enough maybe to get us out of here? (*Miracle in the Andes*, 92–93)

Today, Eduardo Strauch is one of Montevideo's most successful architects. On the mountain, he was one of Marcelo Perez's closest friends. But even he was powerless to shake Marcelo from his growing despondency. Marcelo's need to find reasons had led him to the self-destructive conclusion that he was to blame for this tragedy. As the team captain, he had arranged the match, encouraged everyone to come along, chartered the flight, and hired the pilots. It was his fault. No amount of pleading from Eduardo or others could persuade Marcelo otherwise. When word spread that rescue efforts had ceased, Marcelo's guilt became paralyzing. He grew despondent, and the group became leaderless. But the void was soon filled by something natural and well known to human groups: a clan.

In the absence of Marcelo's leadership, the group increasingly looked to Eduardo's cousin, Adolfo Strauch (called "Fito"). Fito had already proved his resourcefulness by devising a lifesaving water-making system using the plane's reflective materials. He had also fashioned makeshift snowshoes for use by the expeditionaries. When the avalanche hit, Fito kept a level head and directed the others in digging out their buried friends. As a result, in one way or another, many saw themselves as owing their lives to Fito.

Fito was an unlikely leader. He was shy by nature, uncomfortable with public attention or adoration. But he had two close comrades: Eduardo and their mutual cousin, Daniel Fernandez. The three of them were older than most of the other boys (Fito was the youngest of the three), and the others tended to look up to them. But their real strength was one another—that close psychological bond that comes only from kinship. This strength gave them an advantage over all the other potential cliques that formed and fractured over time. They were family, and their bond gave the group a social center of gravity from which desperately needed leadership emerged. When Marcelo died in the avalanche, the "cousins," as they were called, took over, just as kin and clans had done so often before in human evolutionary history.

A Complex Social System

Hunter-gatherers come in two types: egalitarian and complex (for a summary, see Hayden, 2003, 124–125). Egalitarian hunter-gatherers, like the !Kung San (! indicates the "click" sound in their language), are regularly on the move in search of resources. This means that they "own" only what they can carry, which isn't much: hunting equipment, digging sticks, materials for building a shelter, and their vast knowledge of their desert home. Since "property" and "ownership" are all but irrelevant to the !Kung, there is no economic stratification, no rich and poor. Sharing is mandatory. Whoever kills a giraffe cheerfully divides it up among the members of the band. Boasting, claiming special privileges or calling undo attention to oneself is immediately met with ridicule, vicious gossip, and other means of social humiliation and control. Hunter-gatherers can be ruthless in their enforcement of an egalitarian ethic (Boehm, 1999), and accordingly, egalitarian hunter-gatherers have no headman or chief. All decisions must be made by consensus. While some in the band may wield more influence than others, he or she is always careful to do so modestly and evenhandedly.

The aggressive egalitarianism of some hunter-gatherers is easy to understand given the ecological conditions in which they live. Where resources are scarce and widely dispersed, there simply are no opportunities for amassing any sort of "property." To survive, the bands must keep moving. Carrying your possessions night after night is awfully hard on the back. Nor does it make much sense trying to eat an entire giraffe by yourself, should you happen to kill one. Moreover, tomorrow you may kill nothing, and so you'll need the goodwill of the other tribe members to get through the lean times. Share with them now and they'll return the favor later. But not all hunter-gatherers are egalitarian.

In some places, ecological conditions combined with more sophisticated procurement technologies allow hunter-gatherers to become more sedentary and to collect surplus resources. For example, the Tlingit (Native Americans along the northwest coast) use large fishing weirs (traps) to exploit seasonal salmon runs (Oberg, 1973; see also Connors, 2000). In a short time, they can harvest far more fish than the community can eat. Surpluses must be smoked, dried, and preserved. In places where fish runs or herd migrations allow for surpluses to be acquired, hunter-gatherers often stay in one place longer, which leads to social stratification. This stratification usually is based on certain families in the band or tribe who claim a special connection to the resource-rich territories. Perhaps a long-dead relative was the first to make camp there or died there or was led there in a dream. This special connection justifies the family's privileged status within the tribe. This family's ancestors are thought to be supernatural guardians of the territory, and the territory's continued fertility and productivity are essential to the success of the entire tribe. Thus, chiefs, nobles, shamans, and other tribal leaders are rightfully drawn from this elite clan.

Complex hunter-gatherers exist in transegalitarian societies, societies situated between the simple egalitarianism of traditional hunter-gatherers and the multiple strata of settled agriculture (Hayden, 2003, 126). Successfully exploiting a transient but abundant resource like the running salmon requires teamwork and extensive labor. The weirs must be constructed, monitored, and tended while the fish are sliced, smoked, dried, and stored. At the same time, decisions must be made concerning when and where to set the weirs and how long to leave them in place (Connors, 2000). Naturally, the tribal elites make the decisions, and the "commoners" do most of the labor.

Creating surplus resources puts in motion economic forces that often produce social inequalities between the large labor pool producing a surplus for the few powerful clans that control the resource-rich territories. The resource-controlling families can then use the surplus (or part of it) as a trading commodity with neighboring groups to acquire rare and valued items (prestige items). Socioeconomic stratification also brings with it ritual stratification. Elite families or nobles have private rituals designed to fortify and intensify their connection with powerful ancestors (Hayden, 2003, 146).

In some cases, such as among the Chumash of the California coast, these private rituals expand into "secret societies": small councils of elite families who combine religious ritual with political strategizing. Secret societies meet regularly to commune with the spiritual world as well as to plan and direct the tribe's earthly affairs (Hudson & Underhay, 1988). A transegalitarian society thus contains the beginnings of social specialization, in which different strata have different responsibilities in the tribal society: leaders versus followers, chiefs versus braves, owners versus workers, and so on. This includes ritual specialization as well. Among egalitarian hunter-gatherers, shamans are often commonplace. For example, nearly half of male !Kung San are shamans (Katz, 1982). But among complex hunter-gatherers, private rituals require special shamans drawn exclusively from the elite class, whose authority is seen as greater than that of "ordinary" shamans. (Hayden, 2003, 151). Elite power, however, often brings with it severe ritual requirements. Initiations into elite societies can involve deprivation, isolation, and ritually induced psychological and physical pain. For example, as signals of their divine worthiness, Mayan priests and kings were required to engage in ritualized acts of brutal self-mutilation, puncturing their penises or tongues and tugging objects through the open wounds (Schele & Freidel, 1990).

The Evolution of Social Complexity

Chimpanzees, gorillas, bonobos, and orangutans are not egalitarian. Since social hierarchy is a common trait of all great apes (including us), it was very likely inherited from a common ancestor. Thus, it is plausible that our earliest hominin ancestors lived in social hierarchies; yet egalitarianism is a universal trait of simple hunter-gatherers (Erdal & Whiten, 1994). We therefore can infer that something rather extraordinary happened during hominin evolution, that the normal ape social dominance pattern was disrupted somewhere and in its place a more egalitarian society arose. Because this egalitarianism did not last (at least not as a common form of human society), this evolutionary history can be represented as a U-shaped curve, like the following:

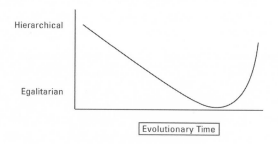

Adapted from Whiten, 1999, 180

Evolutionary time is depicted on the horizontal axis beginning about six million years ago with origin of the hominin branch (extreme left) and moving forward to the current time (extreme right). At first, hominins lived in fairly strict social hierarchies similar to those found among chimpanzees and other social primates. Males challenged one another for dominance, sometimes quite violently, and the winner assumed alpha status with all the reproductive advantages that this entailed. Females also had a social hierarchy, with the more dominant females having greater access to resources for themselves and their offspring.

Over the next few million years, however, this hierarchy was steadily undermined and became more egalitarian. A number of factors contributed to this process. First, even though great apes live in social hierarchies, they also are capable of forming coalitions. A common way for chimpanzees to challenge an alpha male is for two or more subordinate males to team up against him (de Waal, 1982). Thus, the capacity for joint cooperative activity already was present in the earliest hominins some six million years ago, and this capacity provided some of the means necessary to weaken this despotism.

Second, genetic evidence indicates that sometime between 3.5 and 1.2 MBYP, hominins became hairless (Reed, Light, Allen, & Kirshman, 2007; Rogers, Iltis, & Wooding, 2004). Hairlessness combined with bipedal locomotion meant that mothers had to carry their infants with them as they traveled (infants could not cling to their mother's fur as chimpanzee infants do) and that infants would have to be put down while their mothers foraged. This set the stage for what psychologist Michael Tomasello refers to as "obligate cooperative foraging" (Tomasello, 2011). In other words, hominin mothers were very likely forced to forage cooperatively with other females in order to keep watch over their infants, guard against predators, and gather enough resources to feed themselves and their offspring.

Third, stone tool production, which began about 2.6 MBYP, allowed hominins to gain access to a wider array of resources and, over time, to secure larger amounts of resources in the form of scavenged and hunted game. Hunting and scavenging large prey, however, required the cooperation of the group. At first, this may simply have been a group of stick-and-stone-throwing hominin scavengers chasing competitors from a carcass. Since the better-coordinated stick-and-stone-throwers would have had greater success than the less organized ones, cooperation would have steadily taken hold. Then as hominins grew larger and their hunting tools more lethal, scavenging increasingly gave way to cooperative hunting. The strictly enforced egalitarianism of modern hunter-gatherers—depicted as the lowest point in the U shape in the graph—was probably not achieved until about 400,000 years ago when the archaeological record has solid evidence for hominins' big game hunting (Parfitt & Roberts, 1999; Thieme, 2005).

Another important factor contributing to the emergence of egalitarianism (which also is relevant to hunting) is the evolution of hominins' throwing ability. Chimpanzees and other nonhuman primates are not shy about throwing things, although their accuracy is pretty poor relative to humans' (Lawick-Goodall, 1968; Plooij, 1978; Westergaard, Liv, Haynie, & Suomi, 2000). As hominins assumed a more human-like upright stance and stride, their throwing accuracy also increased (Calvin, 1993; Isaac, 1987; Marzke, 1996), and the accuracy with which some hunter-gatherers throw projectile weapons is quite impressive (see Corballis, 2002, 79). The important point about throwing is that once humanlike levels of accuracy were achieved, throwing became a formidable means of social leveling. Indeed, biologist Paul Bingham refers to throwing (or stoning) as a low-cost, potentially lethal punishment-at-a-distance (Bingham, 1999).

For one chimpanzee to challenge another is a high-risk / high-reward venture. The potential reward is obvious: greater status. But the risk—great bodily harm or even death—is often prohibitive. The risk can be lowered somewhat if two chimps team up to challenge a third. But even with a coalition, some bodily risk is involved in any attempt to challenge a despotic authority. It is the great cost associated with punishing despots and bullies that makes egalitarianism hard to establish and maintain. Nonetheless, once our ancestors could accurately throw lethal projectiles at one another, the cost of punishment was lowered significantly. The fact that a bullying alpha male could be taken down at a distance by group of angry stone throwers would make any leader hesitate to force his will on others without first considering the price he might pay. Over time, subordinates' stone-throwing coalitions may not have just "reined in" alpha bullying; they also may have ended it and forced an egalitarian ethos on hominin social groups (Bingham, 1999; Boehm, 1999).

Why did egalitarianism fade? Moreover, note that according to the graph, it did not just fade, it collapsed suddenly in a giant evolutionary heap. The collapse of egalitarianism is yet another example of the reemergence of an ancient way. Unless strong cultural constraints are in place and actively enforced, humans easily and naturally "slide back" to an ancestral form of thinking and behaving, and that is exactly what seems to have happened beginning about 30,000 years ago.

Elaborate Burials

Sungir is an archaeological site located about 150 miles east of Moscow that is famous for an elaborate burial dating around 28,000 YBP. Interred there are an adult male and two children, one male and one female. Each was elaborately adorned with thousands of fine ivory beads, necklaces, pendants, and bracelets. The boy was buried with a belt decorated with 250 fox teeth, while the girl wore a beaded cap and an ivory pin near her throat. Ivory spears, carvings, and other artifacts were also buried with the bodies. Archaeologist Randall White estimated that the number of labor hours necessary to produce the beads alone would have run into the thousands per body (White, 1993). Though less elaborate than the burial at Sungir, the triple burial at Dolní Věstonice in the Czech Republic is of bodies wearing necklaces of ivory and pierced canine teeth, and at Saint-Germain-la-Rivière, France, a young adult female was interred with "imported" and carefully perforated deer-tooth ornaments (Klima, 1988; Vanhaeren & d'Errico, 2005).

Sungir, Dolní Věstonice, and Saint-Germain-la-Rivière are not the oldest burial sites; others date as far back as 100,000 YBP. But these three are among the first to include elaborate grave goods, which in turn provide some of the first evidence of social stratification. It is unlikely that the average tribe member would have been buried with such ceremony. Indeed, among present-day traditional societies, elaborate burials similar to that evidenced at Sungir are reserved for elites who, on death, are presumed to take their place as powerful ancestors (Hayden, 2003, 237–240).

Notable, too, is that many elaborate burial sites, along with many of the sites containing some of the most famous Upper Paleolithic art and artistic artifacts, are located in prominent river valleys where seasonally abundant resources would have been available. Sungir, for example, is located along the Klyazma River, a secondary tributary of the Volga. Lascaux, probably the most famous cave containing art, is located in France's Dordogne valley. Here, not only would the rivers have been important for their fish, but the valleys also probably served as migration routes for grazing animals. Archaeologist Randall White (1985) contends that Dordogne's deeply incised river valleys likely acted as funnels, channeling and constraining large animal migrations, thus making it easier for the resident Cro-Magnon people to harvest large quantities of biomass.

Similar geographic properties can be found in Ukraine, northern Spain, and northern Italy, regions with many Upper Paleolithic sites. Some of the first evidence of storage pits for preserving perishables are dated to around 25,000 YBP and can be found in Eurasia, such as the Kostenki site along the Don River in Russia (Hoffecker, 2011, 123; Klein, 1969). Furthermore, around this period of time we find the first (indirect) evidence for the use of hunting technologies, such as traps, snares, and weirs, capable of procuring surplus quantities of resources (Hoffecker, 2002, 179–184, 239–243).

Stratified social systems were thought to have arisen in association with the agricultural revolution beginning about 10,000 YBP. The evidence just cited, however, suggests that settled agriculture broadened and accelerated a trend that, ten thousand to twenty thousand years earlier, was already emerging in some places. Before agrarianism, in conjunction with increased sedentism, people already were likely to be procuring and storing surpluses in the fertile river valleys of Europe and Eurasia. Here, social complexity and stratification were emerging as well. Settled agriculture gave humans even greater means of producing surplus resources and therefore intensified these trends. In this context, humans "ancestralize" their social world, sliding it back into a state more in line with our primate heritage. Likewise, the social system that ultimately emerged among the Andes survivors was just such a system.

The Mountain Society

The social system that the crash survivors constructed was complex and not egalitarian. Complex social systems arise where an abundance of resources allows for surpluses. In turn, these surpluses must be processed, preserved, and stored. Some members of the society control or own the resources while others process and prepare them. So it was on the mountain. Once the decision had been made to use their dead friends as food, an abundant resource became available, and a stratified society emerged around that resource. This society was as complex as any transegalitarian hunter-gatherer society that anthropologists have documented. It was multitiered with different groups having different rights, responsibilities, and specializations.

THE COUSINS

Marcelo was a "natural" leader in the context of the rugby team, but in the larger context of human evolution, the "cousins" were a more natural form of leadership. Family units make up the core of any hunter-gatherer band, and kinship affiliations compose the social infrastructure of any traditional clan, tribe, or village. When stressed, familial bonds are usually the toughest and most enduring. Thus, it is no surprise that over time, the crash survivors' social system stabilized around a kin group, the Strauch cousins. It wasn't that the cousins were individually any stronger than others but that their closeness gave them a source of inner strength and psychological protection that the others often lacked. This helped them endure the group's shared hardships more successfully. They kept their heads clear and their hopes intact at times when the others were coming unraveled. Undoubtedly, this is one of the adaptive functions that kinship ties have been providing throughout human evolutionary history, so it isn't surprising that others naturally turn to kin and kin groups for support and guidance in difficult times.

As their surname suggests, both Eduardo and Fito were of (paternal) German descent. They were, in fact, double cousins—their fathers were brothers and their mothers were sisters. Daniel Fernandez was related to them on their fathers' side, being the son of their fathers' sister. Even though Eduardo and Fito were as close as brothers, they had distinct personalities. Eduardo was older and physically smaller and had a more cultured, sophisticated demeanor. He had been to Europe; he was calm and thoughtful but passionate about some things, such as architecture, his planned profession. Fito, by contrast, was untraveled, introverted, less decisive about his future, but no less determined or clever than his cousin or anyone else on mountain. At twenty-six, Daniel Fernandez was the oldest of the three, and his calm, steady nature reflected his maturity. When decisions had to made, Fito was usually out in front, with Eduardo and Daniel supporting him. The rest of the group would then usually fall in line behind the three. The cousins exerted an essential centripetal force on the group, keeping it from breaking into a corrosive cadre of bickering cliques. But as with the Mayan nobles, with rank came ritual responsibilities, not all of which were pleasant.

It is one thing to decide to eat human flesh; it is another to actually do it. The meat had to be butchered from the dead bodies, and it had to be sliced into edible portions and processed for consumption, usually by drying it in the sun or occasionally cooking it. The most difficult of these tasks was, of course, the initial butchering of the meat. Even the most tough-minded crash survivors found this nearly impossible. So the cousins assumed this task themselves (sometimes helped by Gustavo Zerbino). The bodies had to be dug out of the snow and thawed sufficiently in the sun to allow cutting. Once the initial butchering was done, large slices of meat would be passed on to a second team who cut them into smaller portions. Once prepared for consumption, the cousins would supervise the rationing of the food to all the survivors. There was a basic ration for all—a handful, about half a pound—but those who worked harder received more, and the expeditionaries received even more than that.

THE LIEUTENANTS

As they gained greater influence and responsibility, the cousins attracted a circle of auxiliaries, slightly younger boys who saw the advantage of ingratiating themselves to the group's authority figures. The auxiliaries or lieutenants tended to stay close to the cousins, convey orders down the line to the rest of the group, do favors, and run little errands for their "superiors." Chief among the lieutenants were three boys: Carlitos Paez, Gustavo Zerbino, and Pedro Algorta.

Carlitos Paez wanted to be an expeditionary—one of those who would try to trek out of the mountains. Because there was some skepticism about his mental toughness, he and some others went on a trial expedition to prove themselves. For Carlitos, it proved to be a harrowing experience from which he almost did not return. Oddly, though, his failure seemed to strengthen him. He made a concerted effort to be more responsible, to work harder, and to behave less like a spoiled brat. He didn't always succeed at this, but his efforts were not unnoticed, and he had another endearing quality: he could make people laugh. Providing comic relief, even unintentionally, became one of Carlitos's important contributions to the group's sanity. Carlitos also served as the group's ritual specialist, leading them nightly in reciting the rosary.

Another of the lieutenants was nineteen-year-old Gustavo Zerbino. Zerbino was a medical student and, as such, was part of the "medical crew" along with Canessa. Being less experienced than Canessa, though, he tended to defer to him on those issues. Zerbino was especially close to Daniel Fernandez and shared with him the responsibility of collecting the money and documents of those who died. Zerbino gained the nickname "the detective," since he often took it upon himself to investigate accusations of minor offenses committed by others. He was a tough fellow—a member of an early expeditionary team who had come back nearly snow-blind. His eyes never fully recovered. At times, Zerbino could be ingratiating to the leaders while bullying the younger boys. Rather a pessimist by nature, when Fito would send him out for weather reports, he would invariable return with bad news, "bitterly cold with a storm brewing," whereas the more optimistic Paez tended give more upbeat assessments, "a little snow but it will clear up soon."

Finally, there was Pedro Algorta, who was somewhat a loner. Introverted and shy, Pedro's best friends were either killed or hurt in the crash. Though relatively uninjured, Pedro suffered a bit of amnesia, completely forgetting that one of his main reasons for going to Santiago was to settle issues with a hoped-for girlfriend. Now he had his sights on returning to Montevideo to *find* a girlfriend. Nonetheless, Pedro was a no-nonsense hard worker and, as such, had caught Fito's attention and goodwill.

THE MEDICAL CREW

The medical crew was composed of two medical students: Canessa and Zerbino, along with Liliana Methol. Liliana's husband, Javier, was a cousin of Panchito Abal, one of the Old Christian ruggers (Panchito was killed in the crash). But the couple's reason for flying to Santiago was more personal: a belated celebration of their twelfth wedding anniversary. In their absence, they had left their four children in the care of Liliana's parents. Ever the mother, the thirty-five year-old Liliana easily adopted the team, especially the younger ones, as her brood, comforting and chiding them when needed. The mountain, however, did not want a feminine nurturing hand. Liliana was killed in the avalanche that struck on October 29, so the doctors were left without a nurse and the boys became orphans.

WORKERS AND PARASITES

Marcelo created a cleanup crew shortly after the crash. One of their chief responsibilities was to clear out the fuselage as much as possible so the survivors could sleep there, protected somewhat from the bitter cold. Once Fito Strauch's water-making system was in place, tending to it became another of the workers' tasks. This division of labor worked fairly well—the stronger, able-bodied workers cleaned up the cabin, while the weak and more badly injured collected water. The work crews were primarily composed of the youngest boys, who filled the lowest rung in the mountain social system. Originally, Marcelo had put Gustavo "Coco" Nicolich in charge of the work crews. Coco put a good deal of energy into keeping up the younger boys' spirits by having them play games and by reciting the rosary nightly with Carlitos. The avalanche, however, also claimed Coco.

Some of the workers drifted into complacency, spending most of the day sitting in the snow, smoking cigarettes, chatting and complaining. They became the community's parasites, and occasionally aroused the resentment of the other group members. There even was talk of not feeding them, but that never happened. Instead, the parasites simply got fewer rations and were otherwise tolerated. They also were another occasional source of comic relief.

Shortly after the crash, Bobby François extracted himself from the plane, plopped down on a suitcase, lit a cigarette, and announced to Carlitos, "We've had it!" That pretty much remained his attitude throughout the entire ordeal. Yet Nando Parrado tells us that Bobby Francçois's "forthright, unapologetic, almost cheerful refusal to fight for his life somehow charmed us all" (87–88). In the mountain society, as likely is true of our ancestral past, no single individual was so valuable as to jeopardize the group's functioning. Traditional societies often ostracize or verbally abuse freeloaders to bring them into line, and the Andes survivors occasionally did this. But they had little time or energy to expend trying to "reform" the parasites. If all they could offer the group was a bit of entertainment, then the group would take it and leave them alone.

THE EXPEDITIONARIES

Their early sorties from the crash site taught the survivors a tough lesson: the high altitude, severe cold, deep snow, and their weakened condition combined to make traveling from the plane a life-threatening proposition. For them to have a realistic chance of trekking out of the mountains to get help, careful planning and preparation would be required. Accordingly, they decided that they should select the most able, both physically and mentally, of the group, and they would try to strengthen and outfit them for a difficult challenge. The expeditionaries would be a separate, warrior caste for whom special prayers were offered and on whom special privileges would be bestowed. They would be exempt from work detail, allowed extra rations, and given the best places to sleep. They stood somewhat apart from the cousins' rule and, as such, provided a useful counterweight to their authority. Neither the cousins nor the expeditionaries could become too despotic, as each checked the others' influence. In the end, four expeditionaries were chosen: Fernando "Nando" Parrado, Roberto Canessa, Antonio "Tintin" Vizintin, and Numa Turacatti.

Moments before the plane crashed, Panchito Abal asked Nando Parrado to switch seats with him so that he could get a better look at the Andes. It was a fateful one-and-a-half-foot transposition. Panchito, Parrado's best friend, was killed in the crash. Parrado was knocked into a coma and, at first, was given up for dead. But during their first night in the crumpled fuselage, Diego Storm, another medical student, spied something in Nando's face that suggested life, and he moved him closer to the survivors to help keep him warm. Over the next three days, Nando slowly roused from his coma to learn that his mother and best friend were dead and that his sister was badly injured (she died a few days later). Nando recovered quickly, and his patience, compassion, and strength made him one of the most respected and best-loved members of the group. There was one point, however, on which he was passionate to the point of near recklessness: to survive they must climb the mountain. If no one would come with him, he would do it alone. At first, most saw Nando's position as suicide. But when the rescue was called off and the avalanche hit, others came around to his way of thinking. He who had been left for dead, rose again in three days to become the leader of the expeditionaries and, in a very real way (albeit not exclusively), their savior.

"And then there was Roberto [Canessa], the brightest, most difficult, most complicated character on the mountain" (*Miracle in the Andes*, 120). Roberto was nicknamed "Muscles," as much a reference to his head as his body. If life were simple, then we could dismiss Roberto Canessa as nothing more than an arrogant bastard who flew into a rage whenever he did not get his way. But life is not simple and Canessa could not be casually dismissed. His temper tantrums were often followed by deeply felt displays of remorse. He often demanded his own way out of a profound sense of responsibility he felt toward others. With only two years of medical school and at the age of only nineteen, Canessa assumed the role of the group's chief medical authority. But as events unfolded, one fact emerged: behind all his bluster and stubbornness, Canessa was as smart and tough as he acted. More than anyone else, Parrado wanted Canessa as his expeditionary partner. If anyone could move a mountain on attitude alone, it was Roberto.

Antonio "Tintin" Vizintin was brave and strong, qualities that made him a good expeditionary. But he could also be self-centered and whiny, with little consideration for others. Among the expeditionaries, "Tintin" was content to take a backseat to Parrado and Canessa, following whatever decisions the two of them made. To the larger group, Tintin was often quarrelsome and bullying. Canessa was one of the few who had some influence over him, possibly because he saw in him something of himself. By contrast, Numa Turcatti was nearly as well liked as Parrado. Turcatti was small amd muscular, but the mountain had taken a toll on him. He found it especially difficult to overcome his aversion to eating human flesh, forcing himself to do so out of a sense of duty to his companions. Furthermore, despite his determination, he had clearly weakened over time. It was Turcatti who suffered a minor leg bruise shortly before the final expedition was scheduled to begin, a bruise that unexpectedly became fatal. Although his death shook the entire group, it had an even more profound effect on his fellow expeditionaries, freeing them from any lingering illusions they might have had about their mountain "home."

The Illusion of Security

SCENE 5: SMILING THROUGH THE HORROR SHOW

[Parrado commenting on the photographs taken during their ordeal] When I see these photographs today, I am amazed, because I see no trace of the terror and depression with which we all struggled. Instead I see us lounging on the snow like college boys at the beach. (*Miracle in the Andes*, 285)

In many ways, the mountain society reflected a complex hunter-gatherer society of the past, except in one critical way. It was not sustainable. It was necessarily a transient society because the abundant resource on which it was based was themselves. Despite the hardships, they were surviving. Conditions were bad, but the battered fuselage did offer protection from the cold and, for the time being at least, they were not starving. They had a routine, and everyone more or less knew his place and responsibilities. Their few attempts at braving the mountain and the snow had nearly killed them. By comparison, their Fairchild home seemed safe. It was only in getting away from that home for a time that they could get an outsider's glimpse at what their life had become. Upon his return to the crash site from a trek to the plane's tail, Parrado began to see the ghastly state to which they had become accustomed—human bones strewn about outside the plane, strips of fat drying in sun, and skulls staring at them from bone piles.

It is a common human practice for the living to return to the site where someone died or was buried. Our closely related hominin cousins, Neanderthals, did so as well. Archaeological evidence, however, points to an interesting difference between *Homo sapiens* and Neanderthals in their *manner of return* to such sites (Wynn & Coolidge, 2012, 111). Neanderthal caves such as Kebara in Israel and Krapina in Croatia contain evidence that the Neanderthals lived among the remains of dead members of their own species. They apparently just pushed them aside and went on with life's daily business. Humans don't do that—not then and not now.

Archaeologist Tom Wynn and psychologist Fred Coolidge suggest that this behavioral difference might point to a deeper cognitive difference in autonoetic thinking. Autonoetic thinking refers to the ability to project oneself backward and forward in time. Contemplating the "what ifs" of life requires autonoetic thinking: "What if I had married Fred instead of Tom?" "What if I leave Tom for Fred?" Envisioning an afterlife also requires autonoetic thinking. Having this mental ability may be one of the reasons why humans recoil from living among or treating casually the dead remains of other humans. "What if they are watching?" "What if it were me?" Neanderthals apparently were not plagued by such discomforting thoughts, making it easier just to get on with the hard grind of Ice-Age survival.

Autonoetic thinking was a double-edged sword on the mountain. It brought revulsion at the intimacy with which they lived among the bones of their dead friends. But it also reminded them of how much worse their lot could be. They had tried escaping the Fairchild before but were turned away each time by the freezing wind, impenetrable snow drifts, and blistering cold. Humans are a remarkably adaptive species—the most adaptive the earth has ever seen. There is no habitat that we have not invaded in some fashion or form. We can get used to anything if the alternative seems worse, and the Andes survivors had gotten used to—even comfortable with—the Fairchild.

Canessa's Delay

As the time drew nearer for the expeditionaries to leave, Canessa found reasons to delay. His sleeping bag was not ready; others needed his medical attention; the radio hinted that a new search might be on. Canessa was never really convinced that climbing the mountain was feasible. He bore the scars of previous expeditions. He understood the hazards all too well. Compared with the relative "safety" of the Fairchild, a trek through the mountains seemed like nothing more than a death sentence. What was harder to see was the creeping death sentence of the Fairchild. Parrado and the cousins tried to convince Canessa that the time to leave was now. But to no effect—no one told Roberto what to do. It was Turcatti who convinced him. He was not supposed to die. He was strong, as strong as they were. A bruise doesn't kill a fighter, a tough guy like Turcatti. But the mountain has it own rules, and death would eventually get them all.

As Turcatti slipped away, a badly shaken Roberto went outside and wilted against the half-buried hulk of the Fairchild, their supposed "home." Parrado pursued him, telling him directly that he and Tintin were leaving the next morning. Was he coming with them? That was the only question.

"Yes," Roberto replied, "It's time to go."

CHAPTER FOUR

Mountain Rituals

SCENE 6: A DAY IN THE LIFE

Despite their woeful state—filthy, freezing, hungry and some hurting badly—a reassuring order slowly emerged among the survivors. Crews were formed, and went about their daily tasks of melting snow, cleaning the cabin, and preparing food. Mid-day was mealtime. As the afternoon sun weakened, they prepared to re-enter the fuselage to escape the descending cold. They filed in by pairs, rotating the line daily so that those sleeping in the coldest spots one night had the warmest spots the next. As they entered, shoes were stowed in the hat-rack to the right. They talked, recited the rosary, and finally slept. The next day, they awoke and did it all again. (*Alive*, 119–120)

Raphael Eschavarren's right calf muscle was nearly torn off in the crash. It was twisted around to the front covering the shin, and even though Zerbino did his best to reset and bind it, it was still a gruesome sight. Eschavarren was in excruciating pain, and over time his condition worsened; the leg became septic and his toes were numb and purple from frostbite. The others did what they could. Canessa rigged up a hammock for him to lie in, and they regularly messaged his feet to try to get the circulation going. Once, after Daniel Fernandez had given him a massage, Eschavarren promised him a lifetime of the finest cheese his family's dairy farm could offer. He had every intention of keeping that promise. No one doubted his fierce determination. His day officially began with his defiant proclamation, "I am Rafael Eschavarren, and I swear I shall return." Often the will to live is the difference between survival and death. But not always. Rafael Eschavarren's fighting spirit eventually succumbed to a body just too far gone.

The challenge facing most of the Fairchild survivors was the inverse of that facing Eschavarren. Their bodies could survive, if only they could keep their spirits from deteriorating in the cold mountain air. The survivors met that challenge in the same way that humans for thousands of years have done—by using ritual for inspiration, emotional cohesion, and as a continual reminder of who they were and what their lives were all about.

The Mountain Grind

Those who weren't already up and about with the morning light were sure to be after Eschavarren's proclamation. Survival required daily effort, and the crews had their assigned tasks—the cabin cleaners, the water makers, the food preparers, and the "doctors" doing rounds. At midday they would line up for their main meal. Everyone had his place in line, and they filed along, cafeteria style, to receive their ration. Overseeing the whole process were, of course, the cousins, who made sure that everyone received his allotted portion: more for the workers, less for the parasites, and most for the expeditionaries. Once, out of frustration at his laziness, the cousins forbade Bobby François from taking his place in the chow line. But instead of protesting or getting to work, he merely accepted his fate, stepping aside sad eyed and pathetic looking. Their bluff having been effectively called, the cousins relented. No matter what, you can't let family starve.

By midafternoon, the sun, and with it, the day's warmth were already fading. Preparations for their nightly return to the fuselage were already under way. As evening closed in, they lined up in pairs in their assigned sleeping order. Everyone had a sleeping partner with whom he would trade shoulder punches, foot rubs, and other means of staving off frostbite. As they entered the plane, each boy removed his shoes and placed them in the overhead rack on the right. Since those who entered first got the warmest sleeping spots at the front of the plane, they rotated the order nightly. The last to enter were the ones assigned to sleep in the coldest spots nearest the piled-up clothes and luggage serving as the plane's back door. As they came in for the night, there was the usual chatter and horseplay that one would expect from active young men. For a time they would talk, but as darkness descended, the calm and quiet grew, and prayer dominated until sleep came.

Each night in the plane, Carlitos Paez served two important functions, one practical and one spiritual. He had arranged with the cousins to be the *tapiador* (the wall builder), meaning that he would always enter last and stack up the sundry stuff used to seal (as best as possible) the plane's aft. Though certainly imperfect, their "back door" helped reduce the effects of the cold night wind whipping through their residence. For this, Carlitos's reward was a permanent sleeping spot nearer the plane's front. His spot was also located near a hole in the plane's fuselage, and thus Carlitos inherited the additional responsibility of emptying the plane's "chamber pot."

Carlitos was also the group's ritual leader. Each night he would lead them in saying the rosary. The rosary involves repetitive calls and responses in which the leader opens with the first part of either the Lord's Prayer or the Hail Mary, and the group responds with the second part. In the darkness, the rhythmic chanting could be hypnotic and was undoubtedly profoundly spiritual for many. In addition, the rosary includes a recitation of the prayer "Hail Holy Queen," which includes the lines "To thee do we cry, poor banished children of Eve / to thee do we send up our sighs, mourning and weeping in this vale of tears." The Catholic schoolboys knew this prayer well, but never before was the plea so intensely personal.

The routines of daily life were punctuated by ceremonies like birthday celebrations. Carlitos himself, along with Bobby François, Nando Parrado, Numa Turcatti, and Pancho Delgado all celebrated birthdays on the mountain. Extra cigarettes, a cake made of snow, and Havana cigars found in the luggage—these and other items served as the gifts. The rosary also was occasionally augmented with special prayers on feast days such as the Immaculate Conception or on the anniversary of a loved one's death. In this way, the group broke their normal routine with special celebrations, which marked time and reminded them of the comforting normalcy of home.

Thirty years after his rescue, Carlitos Paez was asked about the group's daily routines. "Men are creatures of habit," he replied. Routines give people a sense of stability, predictability, and security. They allow humans to control their circumstances, even if that control is more illusory then real (as was the case with the Fairchild survivors). But domestic rituals and routines are much more than this. For humans, rituals and routines are the first layer of the human habitat. Thus Carlitos and his friends were "humanizing" the mountain, carving out a tiny, fragile bit of the mountain's ecosystem capable of supporting, if only briefly, a distinctly human form of living.

Consider Yourself One of the Family

Imagine being invited to dinner by two different families on consecutive evenings. At the first dinner with the first family, the food is left on the stove, and you're invited to simply grab a plate and help yourself. A few people sit at the kitchen table; others are eating and watching television; and teenage daughter stands in the kitchen nibbling a few things while talking on her cell phone. The scene is very relaxed, with people coming and going. You chat with your hosts, eat your fill, and leave.

At the second dinner with the second family, you sit at the dinner table as the food is brought in. You notice that a chair remains empty despite the place set in front of it. Your host explains that this is for the oldest brother, who is off serving in the military. Once the food has been served, you reach for your fork, only to notice that no one else has done so. As you discreetly return your fork to its place, you notice that someone has helped the grandmother to her chair and that only after she has been seated do people begin to eat and converse. When the phone rings, no one answers it.

Our first inclination may be to think that the second family is the one with a "ritualized" dinnertime routine. This may be correct, but not necessarily, because the critical aspect of ritualization is intentionality. Although things could be done in one way, instead they are intentionally done in another. Therefore, the first family may be acting as intentionally as the second family is. In this case, both families' rituals, albeit different, are indicative of their values and priorities.

For example, the second family's respect for familial hierarchy is obvious; the oldest brother's place is still set, and the family doesn't start eating until their grandmother sits down. Furthermore, the group takes precedence over the individual, so phone conversations must wait until dinner is over. By contrast, individual rights (sit where you are comfortable) and responsibilities (get your own food) are highlighted in the first family. The important point is that all families have their own rituals and routines, which serve to highlight and transmit their normative standards.

In addition, as the invited guest to another family's dinner, you have the status of an outsider. As a result, you must be careful, attentive, and on your guard to avoid offending or just looking socially inept. The contrast between your hypervigilance and everyone else's relaxed confidence can sometimes be unnerving. The hosts know their own family "rules"; they're second nature. But you don't, and the effort and attention that you have to expend clearly indicate that you aren't one of them. Although you don't really "belong" here, this has nothing to do with anyone's being inhospitable or unfriendly; it's just that you're not one of the family.

Family Rituals and Routines

Two months of living in the close quarters of half a plane, harassed regularly by discomfort and death, turned the Fairchild survivors into more than just a team or tribe; they became family. Families create bonds through domestic routines and rituals. Repetitive daily activities, broken up by occasional celebrations and ceremonies, make up the core of family life. These activities serve as the social glue holding families together and protecting them from the dissipative forces of narcissism and neglect. Families with strong rituals and routines are more enduring than those without them. Furthermore, strong familial rituals and routines are associated with better social, emotional, and academic outcomes for children. For families in the "real" world, rituals and routines are socially adaptive practices. But for the Fairchild survivors, they were the ties binding them to one another and to life itself.

Psychologists Lisa Schuck and Jayne Bucy define family rituals as "repetitious, highly valued, symbolic social activities that transmit the family's enduring values, attitudes, and goals" (Schuck & Bucy, 1997, 478). These rituals lie at the heart of a family's culture, and they create and reinforce the family members' emotional commitment to one another. Family rituals give family members confidence and a sense of belonging (Bossard & Boll, 1950, 11; Fiese, 2006, 10).

Family rituals are universal and can be categorized into three types: (1) family celebrations, (2) family traditions, and (3) daily rituals (Wolin & Bennett, 1984). Family celebrations are highly organized activities that are fairly standard within a culture and are usually associated with religious and secular holidays (Christmas, Thanksgiving) or rites of passage and transition (baptism, graduation). Family traditions are moderately organized activities that are less culturally specific than celebrations and more idiosyncratic to specific families, such as vacations, visits to relatives, and family reunions. Daily rituals are family interactions that are highly family specific in form but are carried out more frequently than celebrations or traditions. They include repetitive behaviors associated with sleeping (bedtime rituals), eating (dinner rituals), and welcoming and parting.

Routines differ from rituals only by degrees of meaningfulness. In general, routines are more purely instrumental (e.g., doing the dishes, making the bed) and involve activities that family members have to do rather than want to do. Routines and rituals overlap, so a routine can become a ritual if the family gives it special meaning, such as using the "cleaning up after dinner" routine as a way to teach teamwork and sharing responsibilities. Conversely, rituals can become routines when they are perceived as having lost their meaning or become too burdensome, such as when visiting relatives becomes an onerous duty (Boyce, Jensen, James, & Peacock, 1983). Furthermore, some regular activities may have both routine and ritual elements. Mealtime may have routine and not very meaningful actions (setting the table), combined with ritualized and meaningful actions (deciding on seating arrangements and saying grace) (Spagnola & Fiese, 2007).

Teach Your Children Well (Using Routines and Rituals)

According to psychologists Mary Spagnola and Barbara Fiese, family rituals and routines are critical mechanisms for socializing culturally acceptable behavior in young children (Spagnola & Fiese, 2007, 287). Through rituals and routines, children learn how to become responsible, considerate members of a community.

The rituals and routines associated with a simple family meal teach children how to be polite. For example, a "civilized" family meal very likely requires children to sit quietly with others at a table, to take turns in speaking, to say "please" when making requests, to wait for food to be passed to them rather than reaching across the table for it, to use a napkin rather than a shirtsleeve to wipe their face, and numerous other mundane demands that teach patience, respect for others, and self-restraint. Empirical observations have confirmed that mealtime rituals are an important means of transmitting and reinforcing normative social behavior (Goodnow, 1997). One study, for instance, observed eight families with preschool children during mealtimes (Gleason, Perlman, & Greif, 1984) and found an average of 14.5 politeness routines—such things as saying "please" and using napkins. A larger study of more than three hundred families found that most mealtime verbal exchanges were emotionally positive, with 10 percent dealing specifically with proper mealtime behavior (Ramey & Juliusson, 1998). Another 20 percent dealt with family management issues like "who's picking up Tommy tomorrow after school.")

Mealtime rituals have also been found to help teach children the importance of conversational turn taking and the social value of attendance, of simply being present at an activity (Blum-Kulka, 1997; Fiese, 2006, 15–16). Children who are required to attend dinner are more likely to grow up to be parents who go to their kids' important events (school play, basketball game, etc.). Social roles and duties also are highlighted at mealtime. The family hierarchy is often reinforced by seating patterns. The roles and responsibilities of family and community life are enacted and reinforced at mealtimes, with some members preparing and serving food (more often females) and others clearing and cleaning up (Blair & Lichter, 1991; Fiering & Lewis, 1987; Fiese, 2006, 16). Measurable cultural differences have been found in the expectations and treatment of infants and young children during mealtimes. For example, Filipino Americans are highly structured and enforce strict rules of obedience and respect for authority, whereas Caucasian Americans tend to be more tolerant of disruptions, often interpreting them as indicating intelligence and strong will (Martini, 2002).

Other evidence confirms the importance of family rituals and routines to children's social, emotional, and academic development. Although the data are largely correlational, the findings are consistent: a higher frequency of and commitment to family rituals predict more socially competent, confident, and successful children. For example, a five-year longitudinal assessment showed that children whose families were more committed to their domestic rituals had higher scores on standardized tests of academic achievement (Fiese, 2002). The association between commitment to family rituals and higher academic achievement has been replicated among low-income African Americans in both urban and rural settings (Brody & Flor, 1997; Seaton & Taylor, 2003). For boys, family routines were directly related to both academic achievement and self-regulation.

The exposure to the language inherent in many family rituals such as those associated with mealtime or book reading undoubtedly helps with later academic success (Ely, Gleason, MacGibbon, & Zaretsky, 2001). Family rituals also help prepare children for the transition to a structured school environment (Fiese, 2006, 55). Rituals and routines regularly practiced in the home help children appreciate the orderly temporal structure and culturally based behavioral expectations associated with that structure. In other words, children learn that they must take turns, that others may not be able to attend to their specific needs immediately; that they must complete one task before moving on to another; and that they must follow directions and observe temporal contingencies: "You must do your homework before going out to play") (Norton, 1993).

Indeed, the lack of daily routines has been found to be a significant predictor of behavioral problems in children and conduct disorders in adolescents (Keltner, 1990; Keltner, Keltner, & Farren, 1990). When children engage in more daily living routines, parents' reports indicate that they have fewer behavioral problems (Systma, Kelley, & Wymer, 2001). Adolescents whose families have a strong emotional commitment to their rituals have a more positive self-identity and fewer anxiety symptoms, especially under high-risk conditions such as divorce, domestic illness, or substance abuse (Fiese, 1992; Markson & Fiese, 2000). Spanish adolescents referred for mental health services were significantly less likely to have regularly participated in family rituals and celebrations (Compan, Moreno, Ruiz, & Pascual, 2002). Finally, strong family rituals may offer some protection to children in homes with alcoholic parents (Bennett, Wolin, Reiss, & Teitelbaum, 1987; Fiese, 1992).

All this research suggests that family rituals and routines are important for transmitting normative values and behavioral expectations to children. Note, however, that this body of research is largely correlational, demonstrating that those who engage more often in domestic rituals and routines have a variety of positive social, emotional, and intellectual outcomes. Accordingly, we cannot say definitively that rituals *cause* these positive outcomes. The specific causes of these outcomes are undoubtedly multifaceted, with family rituals and routines playing an important but not necessarily exclusive role. At minimum, however, that role appears to be one of providing an unambiguous way for children to observe and internalize norms, values, and behavioral expectations.

Rituals Start Early

Why is it that rituals and routines are so closely connected to positive social, emotional, and intellectual outcomes? To address this question, we have to figure out exactly what ritual is and also to identify the earliest rituals of human life. We may be especially sensitive to the power of rituals and routines because our exposure to them begins almost immediately after birth.

What Is Ritual?

Male baboons use a peculiar ritual to indicate friendly intentions: the scrotum grasp (Smuts & Watanabe, 1990; Whitham & Maestripieri, 2003). One baboon strides up to another using a rapid, straight-legged gait. As he approaches, he looks directly at the other baboon, making affiliative gestures such as smacking his lips, flattening his ears, and narrowing his eyes. The other responds in like fashion, and then after a quick hug, each presents his hindquarters in succession to the other so that he can quickly fondle his jewels. Though odd, this behavior does highlight many of the key features of ritualized actions.

Ritualized actions are formalized, rule-governed, attention-getting behaviors designed to send an important social signal (Armstrong & Wilcox, 2007, 62–76; Bell, 1997; Haiman, 1994; Rappaport, 1999). "Formalized" means an action that has been segregated from a larger behavioral sequence and subsequently is exaggerated, restricted, or stylized in its presentation. Grabbing and ripping at the genitals is common when primates fight; thus, grabbing another's testicles is part of a larger array of fighting behaviors. In the scrotum grasp, this specific fighting behavior has been segregated from the rest and is restricted and stylized in its use. It is now a quick, gentle squeeze, not a violent rip. Formalization can also be seen in the "Mommy, pick me up" signal that toddlers often use. A young child wanting to be picked up often reaches up, grasps at his mother, and begin physically climbing on her. Over time, however, the arm extension act is segregated, and it alone is usually enough to get the mother to bend down and pick up the child. Furthermore, the arm extension act is concurrently stylized or exaggerated (with hands waving to attract attention) but also restricted (eliminating the grasping or climbing motions). Formalization has the effect of isolating the most informative part of a larger behavioral sequence and emphasizing it in order to convey a message more effectively.

"Rule governed" means that the action must follow a prescribed sequence or be executed in a specific way. The scrotum grasp is part of an ordered sequence of specific actions that allow baboons to identify the nature of the signal. You can't just sashay up to another baboon any old way and start grabbing at his private parts. Instead, you must clearly state your intentions up front. The rapid straight-legged gait, direct eye contact, lip smacking, ear flattening, and hugging all must precede the scrotum grasp itself in order to clearly signal the friendly nature of the act. The rule-governed nature of ritualized behaviors helps eliminate ambiguous signals. If you're going to grab someone's testicles—normally an act of aggression—you'd better make clear that you're trying to make friends, not start a fight.

Ritualized behaviors are attention getting. Since they are designed to send an important message, they need to grab the recipient's attention so that he or she does not miss the signal. Although grabbing another's genitals is undoubtedly attention getting, such intimacy is not always needed. More often, repetition and exaggeration are the means by which ritual attracts and holds one's attention. The child wanting to be picked up often repeatedly waves her hands at her mother while making plaintive vocalizations until her mother responds.

There is an important distinction between ritual and ritualized behaviors (Rossano, 2012). "Ritualized behavior" refers to the emancipation, formalization, and rule-governed repetition of elementary gestures. Technically speaking, the baboon's scrotum grasp and the toddler's "pick me up" are ritualized behaviors, not rituals. Ritualized behaviors are widespread in the animal kingdom and are evolutionarily very old. They are nature's solution to the need for cautious, exact, effective communication between members of the same species (Eibl-Eibesfeldt, 1975, 115–124; Tinbergen, 1952).

"Ritual" is a broader term referring to a variety of scripted, ceremonial, and symbolic activities (Bell, 1997). Only humans have rituals. Rituals usually include ritualized behaviors as well as other important features such as sacredness, symbolism, traditionalism, and
performance (Bell, 1997, 138–169). In human rituals, the ritualized
behaviors are embedded in the larger ritual itself. For example, worship at a mosque involves ritualized prayer actions such as bowing
with the palms held upward and kneeling with the forehead intermittently touched to the ground. These actions, however, are surrounded
by ceremonial washing, the symbolism of the mosque and of facing
toward Mecca, and other cultural elements that add to the ritual's
impact.

Baby Rituals (Ritualized Behaviors)

A new mother walks into the nursery expecting her baby to be asleep.
To her surprise, the baby is awake, and as soon as his mother enters,
he opens his mouth wide and emits a happy "Oooouuu." The mother
responds with a cheery "Hellooooo," picks up the baby, and they begin
a happy "turn-taking" exchange of looks, touches, and vocalizations.

MOM (*kisses baby on the forehead*): Look at youuuuu
BABY (*smiles, waves arms*): Heeeee
MOM (*looking into baby's eyes*): I thought you were asleeeeep
BABY (*grabs Mom's nose*): Heeeeee
MOM (*wiggling her nose*): You wanna plaaayyyy?
And on it goes. . . .

About as far from ritual as you could get—right? Wrong. Upon close
examination, the earliest social exchanges between infants and their
mother can be understood as ritualized interactions (Dissanayake,
2000). These interactions have all the essential features of ritualized

behaviors: They are attention-getting, rule-governed, invariantly sequenced, formalized, repetitious social exchanges.

Early "turn-taking" bouts between infants and mothers are typically initiated by an attention-getting signal. These signals usually contain an imitative act (baby smiles and mother smiles in reply, or vice versa) or a "call" in which either the infant directs a vocal signal at the adult or the adult leans toward the infant and vocalizes (Nagy & Molnar, 2004; Reddy, 2008, 52–55; Tronick, Als, & Adamson, 1979). In the preceding example, the baby "called" to his mother by opening his mouth and vocalizing. The mother acknowledged the call by replying in "motherese," or what is more technically referred to as infant-directed speech (IDS). IDS is characterized by high-pitched, slowly paced speech with exaggerated and extended vowel sounds ("Hellooooo youu prettyyy baaabyyy . . ."). These characteristics make it attractive to infants and therefore effective in getting their attention.

Mother-infant interactions involve repetitive behaviors that follow a strict sequence described by developmental psychologist Edward Tronick and his colleagues (Tronick et al., 1979) as

1. Initiation, in which either participant engages the attention of the other.
2. Mutual orientation, in which the infant's initial excitement calms and the caregiver's vocalizations become soothing.
3. Greeting, which is characterized by the infant's smiling and moving his or her limbs and the caregiver's becoming more animated.
4. Play dialogue, in which the infant and the caregiver take turns exchanging sounds and gestures.

The dialogue phase involves mutual turn taking, which has been called "proto-conversation" because of its close resemblance to adult verbal interactions (Keller, Scholmerich, & Eibl-Eibesfeldt, 1988). In other words, when the mother and baby are taking turns exchanging gestures and vocalizations, they are following the same general pattern as that of adults engaged in vocal conversation.

As with adult conversations, the turn-taking proto-conversations between infants and mothers are rule-governed exchanges. Infants not only detect rule violations, but they also protest them, as demonstrated in the famous "still-face" studies (Ross & Lollis, 1987; Tronick, 2003; Tronick, Als, Adamson, Wise, & Brazelton, 1978). In the "still-face," after the mother and infant have established a turn-taking interaction, the mother unexpectedly assumes a nonresponsive, emotionally blank facial expression. Its effect on the infant is dramatic (Hobson, 2004, 36; Reddy, 2008, 73), her mood quickly switching from chattiness to unease. She often looks away and then attempts to reengage the caregiver with a wary smile. When these attempts fail, the infant increasingly withdraws, her agitation mounting the longer the still face remains oriented toward her. There's no question that the still face is distressing to the infant, but can we attribute the distress to the violation of a rule, that "when it's your turn to vocalize or gesture, you should vocalize or gesture"?

That rule violation is the reason for the infant's reaction to the still face has found support in studies using video-mediated interactions between mothers and infants (Murray & Trevarthen, 1986; Nadel, Carchon, Kervella, Marcelli, & Reserbat-Plantey, 1999). In these studies, mothers and infants see and hear each other on live video displays. Then, after a turn-taking interaction between the two is established, a delay is introduced so that the mother's gestures and vocalizations are no longer synchronized with those of her infant. The point of this is to determine whether the infant's distress to the still-face disruption is due to the mother's unresponsiveness or is because of a perceived violation of conversational rules.

In the time-delay manipulation, the mother continues to respond to her infant, with her face remaining oriented toward the infant's, along with smiles and positive vocalizations. Thus, the mother is not emotionally rejecting her infant, but her behavior is no longer properly coordinated with that of her infant—the "turn taking" is disrupted. Infants as young as two months show that they are aware of the change through fewer smiles, more looks away, more closed-mouth expressions, and more general puzzlement and confusion. Thus, it is not just the mother's unresponsiveness or her failure to orient toward the infant that causes the infant's distress. Instead, it is her failure to conform to the expectation of contingent behavioral exchange—an expectation that the infant has already acquired by the age of two months (see also Reddy, 2008, 75–76).

Finally, formalization is seen in the social games between infants and caregivers, such as peek-a-boo, in which restricted, stylized gestures (hands over the eyes, representing a hidden face) are commonplace. Formalization is also seen more generally in that most of the most infant-caregiver interactions over the first few months feature simplified, exaggerated, repetitious movements and facial expressions (called "motionese"; see Brand, Baldwin, & Ashburn, 2002; Schelde & Hertz, 1994). An example is the exaggerated movements used when feeding baby at the table, like the spoonful of applesauce flying in from the east as the father says, "Ooooopen uuuuupp, plane's gotta land!"

As infants mature, both the motionese and infant-directed speech aspects of the infant-caregiver interactions continue to be highly salient when adults are modeling behavioral skills to toddlers and young children. Learning to use utensils, picking up toys, tying shoes, and the many other practical skills that children must acquire are typically demonstrated to them by adults using attention-getting, repetitious, exaggerated gestures and vocalizations (such as "sing-song" melodies) that demonstrate the necessary behavioral actions.

These early interactions between infants and small children are marked by the very same features that define ritualized behaviors: they are formalized, attention getting, repetitious, and rule governed. Ritualized behaviors are the primary mechanism by which we introduce infants and children to the adult social world. Why is ritualized behavior so important in this respect? The answer is that by ritualizing actions, we reduce their ambiguity. We give infants and children a clear and simple message. We are trying to teach them the rules, the norms, of social life.

Using Ritualized Behavior to Teach Social Norms

You have been dropped unexpectedly into the strange land of the melon eaters. They are friendly and welcome you warmly, but their ways are odd and their language is unintelligible. In time, you grow hungry, and to survive you must eat. The land is lush—fruits, nuts, berries of every size and color as far as the eye can see. From the natives' behavior, though, it is clear that some things are edible and delicious, and others are dangerous, poisonous.

A certain melon seems to be a favorite of your hosts. They eat them often, with expressions of delight and relish. But they also do something very bizarre before eating them. Each melon to be consumed is first held out into the bright sun in both hands with arms fully extended. Then slowly the melon is brought to the forehead and gently touched to it three times as the person utters something that sounds like "aboo aboo." There is no mistaking this little pre-melon-eating ritual. More than once, a seemingly perfectly good melon has been passed up when, for some reason, a native could not complete the ritual (cloudy day, injured arm). There is no question: No eating melons unless preceded by the "aboo-aboo" ritual" is a rule taken very seriously here. To you this seems silly. But then again, is it worth taking the chance of insulting your hosts? Or is it possible that the sun somehow detoxifies the melon? With that in mind, you replicate the ritual as best as you can and dive into a delicious melon, and everyone seems quite happy.

In chapter 1, I discussed a series of studies showing that children willingly imitated useless but seemingly ritual actions when those actions were deliberately modeled by adults. So before opening a jar, children would stroke a feather across the top. Or before opening a box, they would wipe a stick across it, front to back, three times. They did this simply because they had seen an adult doing this earlier, and they persisted in doing it even after it became clear that the actions were not physically necessary for opening boxes or jars. These children are assuming that when adults obviously and intentionally perform a certain act, there must be a good reason for it and therefore they should replicate it, even if they don't understand why. But the key to this assumption is intentionality, for other studies show that infants and children will not imitate acts that appear to be unintentional.

For example, Carpenter, Akhtar, and Tomasello (1998) had fourteen- and eighteen-month-olds watch an adult deliberately spin a wheel (while announcing "There!"). Shortly after that, the same adult accidently caught her hand on a lever that illuminated some lights (while announcing "Whoops!"). When given the opportunity later, the infants reproduced the wheel-spinning action but not the lever-pulling-light-illuminating action. Something similar was found among even younger children, one year old, who watched an adult playing with a toy dog that was trying to get into his doggie house (Schwier, van Maanen, Carpenter, & Tomasello, 2006). In one scenario, the dog jumped into his house by going through the chimney, even though the door to the house was wide open. Thus, this act appeared to be freely and intentionally chosen. In a second scenario, the dog entered the house through the chimney, but only after it was found that the door was locked. This act was circumstantially necessitated. Infants replicated the use of the chimney to enter the house in the first scenario significantly more often than in the second scenario.

Finally, suppose a young child watches an adult turn on a light using her head. Will the child assume that heads and not hands should be used to turn on lights? Again, it depends on whether or not the child is convinced that the act was intentional. If the adult performing the act is obviously cold and tightly holding a blanket around her with both hands, then using the head is clearly necessitated by circumstances, and the child will not imitate it. But if the adult's hands are free and she still uses her head, then the act will appear to be intentional, and the child will imitate it (Gerley, Bekkering, & Király, 2002). Furthermore, in the still-face studies, this imitation appears to be rooted in rule learning.

Another study directly addressed the issue of whether children's overimitation (the imitation of causally irrelevant acts) was better explained as stemming from a declarative belief or behavioral procedure. In other words, when the child watches an adult stroking a feather three times over a jar before opening it, does the child learn a declarative belief or rule ("Don't open the jar unless a feather has been stroked three times over the top first"), or does the child learn a behavioral procedure ("Perform the following behaviors to open a jar: stroke a feather over the jar three times, turn the lid counterclockwise until it stops, and pull lid upward").

To address this, Kenward, Karlsson, and Persson (2010) showed four- and five-year-old children how to use a stick to retrieve objects from a box. Before inserting the stick into the box to retrieve items, it was first placed into a dial at the top of the box, and the dial was rotated. When children were given an opportunity to retrieve the items, nearly all of them performed the causally irrelevant act of turning the dial with the stick before using it to retrieve the items (thus demonstrating overimitation). Did they do this because they had learned a rule or because they had learned a behavioral procedure?

To answer this, the experimenters ran a second condition in which before giving the stick to the child, the experimenter placed it into the dial and turned it. What would the child do now? If the child had learned a *rule*, then the rule had been fulfilled (the dial was turned), so she should immediately start using the stick to retrieve items from the box. If the child learned a *behavioral procedure*, then she should still put the stick in the dial, rotate it, and then start retrieving items. Why? Because even though the dial had been turned, the child had not yet performed the *behavioral act* of placing the stick in the dial and turning it. What happened? Most of the children immediately took the stick and used it to retrieve items from the box, demonstrating that they had learned a *declarative belief* or *rule*. They did not know why this rule had to be followed, and most of them understood that rotating the dial was not causally necessary for retrieving items; they knew only that it ought to be done (although not necessarily by them). Other studies have found similar results (Casler & Kelemen, 2005; Williamson, Jaswal, & Meltzoff, 2010). Does this sound a bit like the "aboo-aboo" melon-eating ritual?

The studies reviewed here lead to a straightforward conclusion: infants and children use ritualized behaviors as a means of acquiring rules or norms regarding proper social behavior. They use ritualized behaviors because they are intentional. When someone strokes a feather three times over a jar before opening it, it is hard to see the feather stroking as anything other than deliberate. By their very nature—that is, the fact that they are formalized, repetitious, exaggerated, and rule based—*ritualized behaviors* signal intentionality. In the words of ritual specialist Catherine Bell, they announce, "This is different, deliberate, and significant—pay attention!" (Bell, 1997, 160).

Ritual Against Danger

The Maori *haka* ritual has been made famous by the All Blacks, New Zealand's national rugby team. Before each match, the All Blacks face their opponents and engage in a synchronized display of hand-slapping, feet-stomping, chest-pumping, tongue-wagging, and eye-popping chanting and dancing designed to intimidate their opponents. The All Blacks' version of the *haka* is called *ka-mate*, a war *haka* or *peruperu*. (*Haka* simply means "dance," with numerous forms for different occasions.) Going to war was a serious matter, and before embarking on such a potentially dangerous endeavor, the Maori engaged in ritual. They believed that a well-synchronized prebattle *haka* was a good omen. Anthropologists Pierre Lienard and Pascal Boyer (2006) point out that community rituals are common during times of stress, threat, or danger, which certainly would include war. But *hakas* are not restricted to war; they also are used as a welcome to strangers, as part of a funeral, or as part of various celebrations and ceremonies. All these occasions are, in some way, transitional moments that carry with them varying degrees of stress or uncertainty.

Ritual occasions are often situated at the stress or inflection points of an individual's or a community's life. Christenings, marriages, truces, inaugurations, rain dances, anointings, commencements, initiations, and the like all are highly ritualized affairs, and all mark important and potentially precarious transition points. Though more mundane, even mealtime rituals are associated with a potentially hazardous inflection point in the day. Having gone their separate ways all day long, family members reunite. Some had good days, some not. All of them must eat, but some may be ambivalent about the company or the food selection. Everyone would like the meal to be pleasant, but that is no guarantee. An established behavioral pattern—who does what when and how—may help avoid unwanted domestic unrest.

Another venue full of ritual is an athletic competition. Confronting the prospect of victory's thrill or defeat's agony is inherently stressful, and ritualizing the execution of skilled behaviors is a common strategy for improving performance. Think of the basketball player meticulously repeating the same actions over and over before each free throw or the baseball batter's routine between each pitch (step out of the box, take two swings, adjust cap, take a deep breath, step back in). A number of studies have demonstrated that ritualized behavior (what is technically called a "pre-performance routine") leads to better performance in a number of different sports (Czech et al., 2004; Gayton et al., 1989; Lobmeyer & Wasserman, 1986; Lonsdale & Tam,

2007; Southard & Amos, 1996). For example, basketball players who maintain their established pre-free-throw routines have been found to be significantly more accurate on subsequent free throws compared with those who deviate from their routines (Lonsdale & Tam, 2007). This difference has been attributed to the fact that pre-performance routines serve to eliminate distractions, reduce anxiety, and build confidence by focusing attention on a series of well-rehearsed, productive cues and away from performance-disrupting thoughts (Boutcher & Crews, 1987; Lidor & Singer, 2000).

Whether it is a family wanting a nice dinner, a diver wanting the perfect dive, or a community wanting victory in battle, people facing possible hazards have a big stake in making sure that things go right. It makes sense to ritualize the actions associated with stressful events or transitions—making sure that the message is clear, that the procedures are scripted, that ambiguities and misunderstandings are avoided. Where there is danger, there is ritual.

No one's lives are more fraught with transitions than children's, and there is evidence showing that ritualized behaviors peak at around age five, just when most children are starting school or otherwise making their first moves outside the home environment (Leonard, Goldberger, Rapoport, Cheslow, & Swedo, 1990). Children around this age are often sticklers for precise routines and perfectionists regarding orders and procedures (Zohar & Feltz, 2001). They often see deviations and improvisations as dangerous, similar to how the Maori saw a badly executed *haka* as an invitation to defeat in battle (Evans, Gray, & Leckman, 1999). In a life often marked by new schools, new friends, and constantly evolving roles and responsibilities, it makes sense to identify the social rules most likely to lead to success. Thus, it is not surprising that children both seek out ritualized behavior (in order to learn the rules) and then meticulously enact it (in order to correctly follow those rules).

Ritualized actions highlight the social rules and behavioral norms that guide one through potentially dangerous or precarious situations.

By moving in synchrony in the face of their enemies, Maori warriors (as well as the All Blacks ruggers) signal—to one another and to their foes—their commitment to unity, discipline, bravery, and other "warrior" ethics. Ritualized action embodies the group's values, and those values guide behavior in the face of danger.

Ritual Versus the Mountain

To say that the Fairchild survivors were facing a stressful, hazardous situation is obvious. They were injured, freezing, hungry, and (apparently) trapped. However, as odd as it may seem, these were not the greatest threats to their survival. Their greatest threat was themselves. The severe injuries indeed took their toll, most in the first few days. Many injuries, however, were not immediately life threatening, and the survivors did have some medical supplies and expertise. The cold was tortuous, but huddled in the plane, they could survive it. They had food, but they had to overcome their revulsion against using it, and they had to make it last. The mountain's "direct threats" were difficult, very difficult, but as they proved, they were not insurmountable.

The most precarious thing about their ordeal—the most fragile link in the survival chain—was resisting the "every man for himself" mentality. Could they cooperate, work together, and face the challenges as a community; or would they descend into a self-imploding mob of individual interests? They needed exactly what ritual provides: a set of social norms that guide behavior through hazardous conditions. What were those norms?

LEADERSHIP BY CONSENSUS

Each night in the plane before the recitation of the rosary, there was discussion time. Topics included favorite foods, their families, rugby (of course), and occasionally deeper issues such as theology or political philosophy. This little routine reinforced an important core value of the community; participatory decision making. The cousins were the leaders, but their leadership was predicated on merit and consensus. Fito Strauch did not campaign to be the chief; he was elevated by his actions. At one point, Carlitos and Roy Harley pressed for a public anointing. Fito declined, not because he shrank from responsibility, but because there was no need to formalize what had naturally evolved. Others were going to listen to him and his cousins, regardless of whether or not he carried the title of leader. But it was the *listening* part that was critical to the cousins' leadership. If others were going to listen to them, they would have to make arguments, give reasons, and convince others of a certain course of action. There was discussion, debate, and. finally, a group decision.

SHARED BURDENS

At first, people just slept where they wanted. Quickly, however, it became clear that this particular inflection point in the day was ripe for conflict. The coldest, least desirable sleeping spots were in the back, near the makeshift "backdoor." The wind penetrated the piled-up luggage and clothing so that those farthest in back actually served as part of the windbreak for the others deeper in the fuselage.

In time, they developed a systematic routine for transitioning from the outside world of daytime to nighttime in the dank, dark interior of the Fairchild. As the sun fell behind the mountains and the temperature plummeted, the boys would line up in pairs and file in, with those assigned to sleep in the back entering next to last. Carlitos Paez always entered last and served as the *tapiador*, the wall builder. He would stack the luggage, clothes, and other miscellany that served as their backdoor and then take his assigned spot near the plane's front. The last pair would then take their place by the door. The following night, they would rotate, with those in back moved to the front. This bedtime routine enacted one of the group's core social values, the sharing of the mountain's burdens and discomforts. Under the circumstances, it was more than just a social value, it was a survival value.

PRIVILEGES LINKED WITH RESPONSIBILITIES

Mealtime was another precarious moment in the daily life of the mountain society. Using the bodies of dead friends as food was most unpleasant, and dividing it up fairly to everyone's satisfaction was challenging. As the major architects of mountain social policy, the cousins bore the responsibility of processing and rationing the food. It was they (often assisted by one of their lieutenants, Gustavo Zerbino) who carried out the initial butchering of human bodies. This was a job that even the toughest of the other survivors could not bring themselves to do. Once the meat was initially cut from the bodies, it was passed on to a second team to be prepared for consumption. This team would slice the meat into smaller, more easily eatable pieces and dry it in the sun on the top of the fuselage. The secondary processing was easier, since once detached from the body, the meat could more readily be objectified. For their part, the secondary processors were granted limited pilfering "rights." One out of every five or ten pieces could be consumed on the spot. As long as pilfering did not become too extreme, the authorities were content to turn a blind eye.

Mealtime was at midday and it followed a fairly standard routine. The boys would line up and receive their appropriate ration under the cousins' supervision. The basic ration was about a handful, although the workers could get a little more and the expeditionaries could have as much as they wanted. No one, however, no matter how lazy or burdensome he may have seemed to the others, would be denied food. At the other end of the spectrum, of course, were the expeditionaries, who, compared with the rest, had privileged lives. But it was on them that the group's hopes for rescue rested.

KEEPING THE FAITH

The Fairchild survivors shared a common religious tradition, Roman Catholicism. But within their ranks, their commitment varied. Some— Marcelo Perez, Carlitos, and even the temperamental Canessa—were deeply devout. Others—Parrado, for example—were more laissez-faire Catholics, and a few—Fito Strauch, for instance—could even be called skeptics. Questioning God and even blaming him for their suffering was not an uncommon topic of debate among the survivors. But regardless of their personal feelings, no one protested reciting the nightly rosary. Undoubtedly, this ritual meant different things to different people. For the devout, it was a sincere and heartfelt petition to their God for strength, mercy, and maybe even a miracle. For the more skeptical, it was a means of mental relaxation; something to help them get badly need sleep and something to help preserve their sanity. No one put himself above the ritual. Group solidarity, expressed and reinforced in communal prayer, was more important than any single person's doubts or misgivings about the supernatural.

One night, not long after the avalanche, the group lay wide awake in the pitch-dark hull, menaced by the rumblings of the volcanic Mount Tinguiririca. In their fright, they pushed the rosary beads into Fito's reluctant hands and demanded that he lead them in prayer.[1] He complied, praying with all the fervor of one newly ordained. The rumbling ebbed and finally ceased. Whether faith in God or faith in one another, the critical thing was faith.

WE ARE FAMILY

They fought and they reconciled. They threatened the lazy with starvation, but their hearts softened before they could follow through. They celebrated birthdays with whatever meager gifts they could find or improvise. They ate together and slept together, huddled in pairs against the bitter cold. They prayed for the strength to make it through another day. For seventy-two days, the Fairchild survivors used their rituals and routines to pronounce one unambiguous message to their mountain prison: Our fates are interlocked; we are family.

CHAPTER FIVE

Rituals of Love

SCENE 7: LOVE OVER DEATH

Certainly their quarrels were never serious when compared to the strong bond of their common purpose. Especially when they prayed together at night they felt an almost mystical solidarity, not only among themselves but with God . . . [and] with the friends who had died and whose bodies they were eating to survive. (*Alive*, 193)

Death has an opposite, but the opposite is not mere living. It is not courage or faith or human will. The opposite of death is *love.*" (*Miracle in the Andes*, 201, italics in original)

What He Did for Love

He was popular. He was handsome. And he was the king of England and all its far-flung possessions. But by the end of 1936, he had renounced it all for love. He couldn't live without her, Wallis Simpson, soon to be divorced for the second time, then a scandal in British society.

Edward VIII's was romantic love. But romantic love is not the only kind of love that can compel sacrifices from those afflicted. It was the love of his father that drove Nando Parrado to climb a mountain, enduring all the pain and hardship that that entailed, in order to return to civilization. Love of family, group, or nation can be equally powerful in motivating self-sacrificial behavior. The Fairchild survivors had to cultivate this form of love in order to resist the selfish inclinations that might have destroyed their community and led to disaster. Even though they all took advantage where they could in order to try to improve their individual lot, they never let this weakness overwhelm their commitment to their common cause. Love of the group remained stronger than personal fear.

Running into a burning building to save one's child, dying for one's country, and even suicide bombing all have their roots in a powerful emotional attachment to those we consider "our people": our clan, our fellow group members; the "us" in the us versus them reality of human life (Richerson & Boyd, 2001). A highly effective way of compelling humans to sacrifice themselves for others is to get them to love those others. When members of a family, team, tribe, or nation have such love for one another that they are willing to set aside their personal interests and endure great hardship for the benefit of the group, they can overcome great obstacles and surmount enormous odds. How do group members learn to love one another that much? Common genetics is important, but when the group extends past relatives, the answer again is ritual.

In 2009, Jeremy Ginges, Ian Hansen, and Ara Norenzayan, three psychologists at the University of British Columbia, completed a comprehensive cross-cultural study looking at a phenomenon called "parochial altruism" (Ginges, Hansen, & Norenzayan, 2009). Parochial altruism refers to a self-sacrificial act made on behalf of one's group, an extreme form of which involves aggression or violence against an out-group (Choi & Bowles, 2007). Examples are the current terrorist acts of suicide bombing, the bombing of Allied ships during World War II by Japanese kamikaze pilots, the sacrifice by three hundred

Spartans at Thermopylae, and the many acts by the French resistance during the Nazi occupation or by the members of the African National Congress during apartheid in South Africa.

In the specific case of suicide bombing, many people believe that religious belief is the prime motivator, with the attacker believing that his or her sacrifice will lead to a glorious reward in the afterlife. But another possibility is what Ginges, Hansen, and Norenzayan call "coalitional commitment," the powerful emotional attachment among group members. Indeed, this attachment can be so powerful that a group member may see dying as preferable to letting the others down. Frequent participation in group ritual can be an effective mechanism for creating these strong emotional attachments.

Ginges, Hansen, and Norenzayan measured belief and coalitional commitment through surveys of people's self-reported frequency of both praying (an indicator of religious belief) and attending worship services (an indicator of ritual participation), and they compared these measures with their support for suicide attacks (or acts of parochial altruism in general). The surveys were conducted among Palestinian Muslims (residents of both the West Bank and Gaza), Indonesian Muslims, Mexican Catholics, British Protestants, members of the Russian Orthodox Church, Israeli Jews, and Indian Hindus. In every sample surveyed, it was attendance at worship services, and not frequency of prayer, that predicted support for suicide attacks. Indeed, in at least one subsample (Indonesian Muslims), frequency of prayer was negatively correlated with support for parochial altruism; that is, the more devoted Muslims were more likely to oppose suicide attacks. These results support the notion that frequent participation in group rituals builds the kind of intragroup emotional commitment that can lead to self-sacrificial acts.

To further validate their findings, these researchers conducted an experimental manipulation with Jewish settlers living in either Gaza or the West Bank. The manipulation entailed "priming," in which a subtle, often subliminal, reminder of a particular concept temporarily increases the concept's influence on a person's attitude or behavior. For example, previous priming studies have found that if people are reminded of a particular religious (God, spirit) or legal (court, police) concept, they will act more generously in subsequent economic games (Shariff & Norenzayan, 2007).

In this priming study, half the settlers were randomly assigned to a synagogue prime in which they were subtly reminded of worshipping at a synagogue, and the other half were exposed to a prayer prime in which they were reminded of praying to God. After this, subjects were asked whether they regarded Baruch Goldstein's 1994 attack on a mosque to be "extremely heroic." Significantly more subjects receiving the synagogue prime (23 percent) affirmed this statement than did those receiving the prayer prime (6 percent). This suggests that when people have group ritual behavior on their minds, they tend to be more sympathetic to self-sacrificial acts of out-group aggression. This supports the hypothesis that forming group identities and emotionally binding people to those identities are important driving forces behind this behavior. While religious ritual is a highly effective group-bonding mechanism, it is not unique in this respect. Fraternities, military services, and social/political movements all use the same basic principles and processes.

In-Group Love or Out-Group Hate?

Can we be sure that this self-sacrificial behavior is really motivated by love of one's group? Might it not be just hatred of members of the out-group? Two lines of experimental evidence are relevant here, and both point more to in-group love than to out-group hate as the important motivating factor. First, a number of studies provide evidence that ritualized group actions can produce greater cooperation, compassion, and altruism among group members (McNeill, 1995; Valdesolo & De-Steno, 2011; Wiltermuth & Heath, 2009).

People who move together emotionally bond together. Historian William McNeill wrote extensively about how such rhythmic group actions as communal dancing, chanting, or marching produce a euphoric mental state and a "muscular bonding" among the participants. He describes his own experience of participating in military drills as creating a "euphoric fellow-feeling . . . [a] strange sense of personal enlargement, a sort of swelling out, becoming bigger than life, thanks to participation in collective ritual" (McNeill, 1995, 2). He notes that numerous rebellious or countercultural movements use dance or similar group-coordinated actions as a means of creating enthusiasm and loyalty. Examples are the Beni dancers in colonial Tanganyika, the liberty tree dances in revolutionary France, the 1960s hippie dances in the United States, the ghost dance movement among Native Americans, and the Melanesian cargo cult dance (McNeill 1995, 59–65).

Experiments confirm that moving together enhances liking, cooperativeness, and compassion (Hove & Risen, 2009; Valdesolo & De-Steno, 2011; Wiltermuth & Heath, 2009) and can also increase the endurance and pain tolerance necessary for achieving difficult collective goals (Cohen, Ejsmond-Frey, Knight, & Dunbar, 2010). For example, Wiltermuth and Heath (2009) experimentally manipulated ritual-like activity and then measured prosocial behavior. They had subjects engage in synchronized motor movements (walking in step, singing in synchrony, singing and moving in synchrony), nonsynchronized movements (walking at one's own pace, singing and moving individually), or no movements at all. Later, all the subjects played an economic game in which they could extend varying levels of trust and cooperation to other players. Those subjects who engaged in synchronized movements were found to be more trusting and cooperative than those in either the nonsynchronous movement or the no movement conditions.

Why do people who move together trust each more? One possibility is that moving in synch produces greater liking, and indeed, there is evidence supporting this (e.g., Hove & Risen, 2009). When someone matches another's actions during conversation (crossing one's legs or nodding one's head when the other does), then likability ratings tend to go up. Thus, if you want to be received favorably in a job interview or convince someone to go out on date with you, you should match that person's movements while talking with him or her. But there may be more to it than this. Psychologists Piercarlo Valdesolo and David DeSteno found that synchronized movements increased both likability and perceived similarity (Valdesolo & DeSteno, 2011). But their analysis also suggested that perceived similarity, more than likability, was responsible for generating feelings of compassion that motivated acts of altruism toward synchronous others.

The participants in Valdesolo and DeSteno's study engaged in either synchronous or asynchronous finger tapping. Later, those who tapped synchronously reported having a greater liking for one another and a greater perceived similarity. In another part of the study, the participants were made aware that some of the study participants (the "victims") had been assigned an onerous task and that they could relieve them of some of that task (i.e., they could behave altruistically toward them). Participants expressed significantly more compassion for those with whom they had earlier moved in synchrony and they accordingly extended greater altruism toward them. A path analysis indicated that it was the perceived similarity factor that played the more critical role (compared with the likability factor) in producing feelings of compassion and altruistic behavior.

Synchronized movements can also increase pain tolerance, thereby aiding in the endurance necessary to achieve group goals. Cohen and colleagues (2010) measured pain thresholds in rowers who either rowed alone or in synchrony with others. Pain thresholds were significantly elevated for those who rowed in synchrony with others. Exercise like rowing is known to cause the release of endorphins, often producing a natural high (Harbach et al., 2000; Howlett et al., 1984). Cohen and colleagues' study indicates that endorphin release may also be associated specifically with group synchrony. Along with the release of endogenous opiates, synchronized activities lead to "linkage," or correlated patterns of activation and deactivation of the autonomic nervous system (ANS), such as breathing patterns and heart rhythms. Linkage thus has been hypothesized to be an important physiological basis for empathy (Frecska & Kulcsar, 1989; Levenson, 2003).

Research on cooperative games is the second line of experimental evidence indicating that in-group love is more powerful than out-group hate in promoting self-sacrificial acts. These studies show that in-group bias and out-group animosity are separable constructs, with in-group bias being driven more by love of the in-group than by hate of the out-group (Yamagishi, 2007; Yamagishi, Jin, & Kiyonari, 1999; Yamagishi & Mifune, 2009). In these studies, the subjects were placed into game situations in which they were partnered with someone who was either an in-group member or an out-group member or whose group membership was unknown ("strangers"). The groups were "minimal" groups, formed immediately before the experiment and based simply on the person's stated preference for the work of one artist (Klee) or another (Kandanski). The game was played as follows:

> There are two players. Each player gets a sum of money (say $10). Each player decides how much to give to his or her partner. Each player keeps what is not transferred to the partner. The amount transferred to the partner is doubled by the experimenter. Each player's total is composed of what he or she kept plus the amount the partner transferred to him or her (doubled by the experimenter). Thus, the game comes down to trust. If you can trust your partner to reciprocate your generosity, then both of you can reap a greater end profit compared with simply keeping all of your initially allotted money.
>
> For example, suppose I don't trust my partner. I transfer zero to my partner and keep my $10. If my partner transfers any nonzero

amount to me, I will come out ahead. If not, at least I won't lose anything. But what if I trust my partner? I transfer all $10, and my partner does the same. The ten is doubled, and we both end up with $20, twice what we started with. A player's generosity (i.e., the amount transferred to the partner) is considered the index of cooperation.

The first of two important findings is that the subjects were more generous with in-group members than out-group members. In general, the reason for this appears to be an expectation of mutual in-group generosity and reciprocity (something not expected of strangers or out-group members). Failing to live up to this expectation is seen to harm one's reputation. There is some indication, however, that males may be cooperative with fellow group members even when they don't expect reciprocity, suggesting that they may sometimes use cooperation simply as a display of group solidarity (Yamagishi & Mifune, 2009).

The second important finding is that players were no less cooperative with out-group partners than they were with strangers (those whose group status is unknown). Thus, there is no negative bias specifically targeted at out-group members. Instead, they are approached with the same degree of caution as a stranger, that is, the same, generally lower, level of cooperation. Furthermore, these studies found no negative correlation between in-group cooperativeness and out-group uncooperativeness. In other words, those who were more positively biased toward the in-group were not necessarily more hostile to the out-group, as might be expected if in-group love and out-group hate were linked (i.e., the more he loves "us," the more he will hate "them"). Neither the stranger nor the out-group member is immediately treated spitefully. Instead, the default orientation toward strangers or out-group members is not hate but caution. Caution could develop into hate, however, as competition and distrust mount.

Prayer Against the Mountain

The research just reviewed leads to two important conclusions, that (1) love of the group can motivate individuals to engage in self-sacrificial behavior and (2) group rituals can build group love. The nightly rosary was the Fairchild survivors' most powerful group ritual.

The call-and-response prayer in the cold dark hull of the crumpled fuselage was undoubtedly an emotionally powerful and mentally hypnotic experience. The repetitive cadence forced the participants to synchronize their breathing, vocalizations, and other motor and autonomic functions. Indeed, lab studies of rosary recitation have found that it slows and synchronizes breathing patterns and heart rhythms while also inducing feelings of well-being (Anastasi & Newberg, 2008; Bernardi et al., 2001). Thus, the survivors' nightly prayer ritual created the same conditions found in laboratory studies to lead to the release of endogenous opiates (endorphins) and the ANS linkage thought to produce both pain relief and group empathy. Furthermore, the meditative state achieved during this ritual may have led to the release of other health-enhancing neurochemicals such as arginine vasopressin (AVP). AVP levels have been shown to be higher in meditators than in nonmeditators and has been associated with decreased self-perceived fatigue and increased positive affect (O'Halloran et al., 1985; Pietrowsky, Braun, Fehm, Pauschinger, & Born, 1991).

To survive, the Fairchild survivors had to elevate group love over personal fear, and the rosary created the conditions for doing so. The sense of solidarity strengthened and renewed nightly in group prayer was augmented by the increased sense of personal well-being and elevated pain tolerance (however slight these might have been, given the conditions) that likely also occurred during the ritual practice.

One objection to this analysis is that it entirely discounts the rosary's religious aspect. In other words, a physiological/psychology explanation for the effects of ritual prayer trivializes its religious significance. One response to this, of course, is that explaining the physical basis of something does not necessarily explain it away. But more relevant to this concern is that religious belief itself has been shown to play a role in pain tolerance.

Researchers at the University of Oxford compared the pain tolerance of devout Catholics and nonbelievers while each contemplated a religious or nonreligious image (Wiech et al., 2008). Both the religious and the nonreligious subjects had been tested earlier for equivalent levels of pain sensitivity. While shock was administered, the subjects looked at either a religious image—*Vergine Annunciate* by Sassoferrato (the Virgin Mary praying)—or a similar, nonreligious image—*Lady with an Ermine* by Leonardo da Vinci. A significant increase in pain threshold was found only in the Catholic subjects when viewing the religious image. In other words, when the Catholic subjects looked at the image of the Virgin Mary, they could endure higher levels of shock than when they looked at the *Lady*. Furthermore, when viewing the Virgin Mary, they could endure more intense shock than the nonreligious subjects could, regardless of which picture they viewed.

The researchers also monitored the subjects' brain activity while they were being shocked. When the Catholic subjects viewed the Virgin Mary, they showed increased activity in an area of the brain known to be involved in the evaluation and modulation of pain (the right ventrolateral prefrontal cortex, rVLPFC). Similar activation was not found in the nonreligious subjects.

These results confirmed the hypothesis that the sense of peace and compassion that often accompanies religious contemplation serves as a powerful force in pain reduction. As they recited the rosary, the Fairchild survivors pleaded with the Virgin Mary, no fewer than fifty times, for help and intercession (the rosary prayer entails fifty repetitions of the Hail Mary). This research suggests that for the devout at least, that plea was very likely answered.[1]

The Urgency of Ritual

Nando Parrado rejected any characterization of him and his fellow survivors as saints (see *Miracle in the Andes*, 275). They all were prone to selfishness, and they all sought to better their own circumstances, sometimes at the expense of others. What saved them, he contended, was that their natural self-interest was ultimately outweighed by their collective concern for one another. But how does a society keep a "natural self-interest" from outweighing a "collective concern for one other"?

For decades, the English ethnomusicologist John Blacking studied the music making of the South African Venda people. (His work is standard reading for those interested in the origins and social functions of music.) He saw the Venda as a people strongly committed to social cooperation and egalitarianism (he may have idealized them somewhat in this regard), and he makes an intriguing observation about Venda communal music making in his book *How Musical Is Man?* (1973, 101). Communal dancing and music making, he observed, actually increased during times of plenty than during leaner times. He argued that when individuals and cliques were most tempted to pursue their own interests, communal rituals took on their greatest urgency. It's when we think we don't need or can't afford the group that we are most in need of reminders of how important the group is.

As we saw in chapter 4, rituals often are most productive during social "stress." When we need to ensure that social interactions or transitions will proceed smoothly without dangerous misunderstandings or ruinous complications, ritual is often the answer. For the Venda, times of plenty could be threatening, as the opportunity for selfish indulgence might prove irresistible for some, which could fray the social fabric. Ritual singing and dancing, though, reinforced communal norms and dissipated divisive urges.

There was nothing particularly special about the sixteen young men who survived the ordeal of UAF flight 571. They were susceptible to all the faults and fears typical of their age and upbringing. But they did have a common concern—survival—that demanded collective effort and the restriction of potentially fatal self-interests. Weariness, fear, and hopelessness were constantly eroding their communal ethic. To survive, the members of a society in such constant peril had to find ways of regularly reenergizing their social commitment, and that is the function of communal ritual.

The Evolution of Ritual

Primates are highly social animals with a wide repertoire of ritualized actions for regulating their social life. The baboons' "scrotum grasp" is one example. When foraging parties of chimpanzees reunite, they greet one another with an array of welcoming rituals: mutual embraces, kisses, grooming, and group pant-hooting (Goodall, 1986). Chimpanzee combatants express their desire for reconciliation by using submissive bows and the extended-hand begging gesture (on the part of the loser) and embraces and kisses (on the part of the winner) (de Waal, 1990).

Human rituals differ from those of other primates in two important ways. (1) Humans are able to move together as a group in coordinated and synchronized ways. That is, humans dance, sway, march, and chant together, whereas monkeys, apes, and other primates don't. (2) Humans consciously construct ritualized gestures and link them into behavioral sequences, which other primates generally do not do. This conscious ritualization segregates elemental gestures from their instrumental ends and formalizes them into meaningful social signals. For example, washing a window requires several motor movements, one of which is a repetitive up-and-down or circular hand motion. This motion has an instrumental end: getting the window clean. In ritual washing, however, this motion is intentionally isolated, formalized, and detached from its instrumental end. The purpose of ritual washing is not to clean the object but to signal its sacredness. In addition, the ritualized washing motions are often combined with other ritualized actions to create an entire ritual sequence. The object is held in a certain way, ritually washed, reverently returned to its place, and the washing cloth is folded in a prescribed manner, and so on. All this makes up the ritual-washing sequence. In this way, the ritual gestures become ends themselves rather than the means to an end.

Learning to Move Together

If symmetry is beautiful, then male fiddler crabs are one of nature's ugliest creatures. One claw is normal sized while the other is grotesquely huge, sometimes as much as one-third of his total body weight. Although the claw is useless for feeding, it is good for fighting and even better for attracting females. When a female approaches, a male begins swinging his claw, moderately if he's alone but more vigorously if other males are in the vicinity. A curious thing happens when a female encounters a group of males: they surround her and begin swinging their claws in synchrony. This is not a crab version of the human "wave," which is intentionally cooperative, as people are trying to synchronize and coordinate their motor movements in order to send some signal. Instead, fiddler crabs are actually competing with one another to attract the female's attention. In a study using robotic fiddler claws, Reaney and colleagues (Reaney et al., 2008) exposed female fiddler crabs to both synchronously and asynchronously waving claws. They showed no preference for synchrony. Rather, the females actually preferred the member of the waving group that appeared to be the leader. Synchrony, Reaney and colleagues concluded, arises as a by-product of males each trying to convince the female that he's the group's wave leader.

Vying for leadership is not the only reason for natural synchrony. It frequently emerges when physical or biological systems are in proximity to each other. In the mid-seventeenth century, Dutch physicist Christiaan Huygens found that by mounting two similarly designed pendulum clocks next to each other on the wall, their pendulums would, over time, synchronize. But if one were moved across the room, the synchrony usually would be lost. In this instance, physical proximity allowed perturbations to propagate through the air or wall from one clock to another, entraining their rhythms. Something similar also appears to happen with the menstrual cycles of women who share living quarters like college dormitories or prisons. Here the communication channel is likely chemical rather than physical, pheromones passed from one female to another, providing information about the stage of the female's cycle (Stern & McClintock, 1998; but see also Yang & Shank, 2006).

Other cases of synchrony have been attributed to the "precedence" effect (Litovsky, Colburn, Yost, & Guzman, 1999; Wallach, Newman, & Rosenzweig, 1949), which most commonly occurs when one animal vocalizes and another matches that vocalization either to intensify its signal strength or to mask the initial signal. In either case, competition, and not cooperation, is the motivation. A group of males may synchronize their signal calling as a means of outcalling another group (Buck & Buck, 1978; Wells, 1977), or the calling may become synchronized as the males each attempt to mask the others' calls. This appears to be the explanation for much of the synchrony found in creatures like fireflies (which sometimes "flash" together) or frogs and katydids (which, respectively, croak and click together; Buck, 1988; Greenfield & Roizen, 1993). Dolphins may naturally swim and surface synchronically because of physical factors affecting aquatic navigation (e.g., minimizing drag). There also is some indication that it may serve as a mechanism for reducing tension and increasing cooperation among alliance members (Connor, Smolker, & Bejder, 2006).

From these studies, we can see that physical and biological systems often fall naturally into synchrony when they are in close proximity to one another and/or when they are competing with one another. What is rare (maybe nonexistent) outside the human world is intentionally cooperative synchrony, in which individuals consciously regulate their motor movements in order to bond emotionally with one another. Even higher primates rarely move together rhythmically.

Some examples, however, suggest that our primate relatives are tantalizingly close to being able to sing and dance together. For example, gibbons, small tree-living apes found in Indonesia, are unique among apes in that they form (relatively) monogamous mating pairs. A gibbon pair often engages in extended call-and-response duets that may serve to strengthen the stability of their bond (Geissmann, 2000; but see Brockelman, 1984), as well as to define and defend their mutual territory (Farabaugh, 1982; Mitani, 1985). Similarly, both chimpanzees and bonobos engage in pant-hoot chorusing, in which they closely match their vocal calls (de Waal, 1988; Mitani & Brandt, 1994). These appear to be associated with territorial defense and celebrations that occur when foraging groups reunite, suggesting that positive social emotions are associated with them.

Despite these examples of vocal coordination and synchronization, there are virtually no examples of *moving* in synchrony. Indeed, a number of researchers have noted that among primates, only humans seem capable of entraining their movements to a shared rhythm (Atran, 2002, 171; Brown, Merker, & Wallin, 2000; Williams, 1967). The one possible account of group-coordinated movement among chimpanzees comes from Wolfgang Kohler (1927, 314–316). At his research station at Tenerife, on the Canary Islands, Kohler found that while playing, a group of chimpanzees began to "march in an orderly fashion in a single file around and around the post. . . . a rough approximate rhythm develop[ed] and they tend[ed] to keep time with one another." Kohler remarked that nothing he had seen before from the chimps so strongly reminded him of the dancing of some primitive tribes. To some degree, however, it appears that the chimpanzees' primitive dance benefited from emulating humans. Kohler went on to describe how he could encourage their dance by stamping his foot rhythmically and how they typically halted their dance (with great disappointment) when his stamping ceased.

Thus, the ability to rhythmically coordinate group-level movements appears to be just beyond the abilities of our closest animal relatives. Given that chimpanzees can almost move in synchrony and given how easily synchrony appears to arise in nature, it seems that moving together in dance and song was probably not very difficult for our hominin ancestors to master. Probably the critical element was developing the conscious motor control necessary to regulate movements in coordination with others. For this, toolmaking may have provided the essential selection pressure.

Stone Tools and the Evolution of Motor Control

Dancing together requires that we be capable of regulating our own movement patterns in accordance with another person's movements. Put more simply, I must get my body to do what I see you doing. This exercise in perceptual motor control appears to be just beyond what chimpanzees and other nonhuman apes can achieve. Our hominin ancestors, however, did manage to achieve this ability, and stone tool manufacture is very likely reason for it. Considerable evidence shows that the use of stone tools brought with it an unprecedented expansion of hominin motor control. As mentioned earlier, there is evidence that a humanlike bipedal stride was in place by 3.2 MBYP. The relevance of this to tool manufacture is that uprightness brings with it increasing opportunities for the visual guidance of hand movements, making those movements available for conscious awareness and control. Although the first evidence of stone tools does not emerge until about 2.6 MBYP, it is reasonable to assume that bipedal hominins were making and using tools constructed from perishable materials long before this.

Chimpanzees commonly fashion tools from branches, leaves, and other organic matter for hunting, gathering, and cleaning purposes (Goodall, 1986; Pruetz & Bertolani, 2007). This suggests that our earliest hominin ancestors also were doing so as far back as 4 or 5 MBYP. In contrast, though, they were doing so while standing upright, visually inspecting and guiding their hand movements. Furthermore, the upright stance brought with it skeletal modifications to the hands and arms, which likely improved their manipulative abilities. In turn, these preconditions very likely set the stage for the earliest stone tool industry, the Oldowan, dating to around 2.6 MBYP and originally discovered in northern Kenya (Semaw et al., 1997). Although these tools are nothing more than sharp flakes struck from a pebble core, studies reveal that the percussive striking motion used to create Oldowan flakes has unique gestural qualities.

THE OLDOWAN AND THE EXPANSION OF MOTOR CONTROL

Wild apes do not make or use tools comparable to Oldowan tools (Marchant & McGrew, 2005; Toth & Schick, 2009), although captive apes have been taught to make Oldowan-like tools. When making them, however, they do not use the same percussive striking gesture used by hominins (Toth, Schick, Savage-Rumbaugh, Sevick, & Rumbaugh, 1993). The Oldowan knappers used a technique known as a "conchoidal fracture," in which you take a large stone and hold it facing upward in your left hand (assuming you are right-handed; if you are left-handed, just reverse everything). This is your core stone, the stone from which you are going to detach sharp flakes. In your right hand, grip a smaller, denser stone. This is your hammer stone, the stone you will strike against the core to detach a flake. Hold the core stone comfortably out in front of you and the hammer stone above it high enough to generate a striking force that you think will cause a break in the core. Now, when you strike the core with the hammer, make contact near, but not on, the edge of the core at an angle no less than 90 degrees (Pelegrin, 2005, 25). That's a conchoidal fracture, also called the percussion technique, as shown in the following figure.

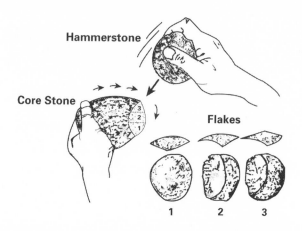

Even though this is not too complicated, it represents a level of motor control beyond that of extant nonhuman apes. How do we know this? In a few studies, nonhuman apes have been taught to make stone tools. The most famous was of the language-trained bonobo Kanzi. Kanzi successfully produced sharp flakes, but he did not use conchoidal fracture. Instead, he employed something that researchers term "split breaking," throwing a core onto the ground or against a wall in order to break off flakes. Split breaking is more violent, less precise, and reliable in producing usable flakes.

Whether conchoidal fracture represents an important cognitive advance is debatable, although it is a significant advance in motor control and coordination. Wild chimpanzees often crack nuts by striking a nut with a rock to break open the shell. More than likely, Oldowan knapping movements are an evolutionary extension of this nut-cracking motion (Marchant & McGrew, 2005), although studies comparing chimpanzees' nut cracking with percussion flaking reveal the greater complexity of the latter. Oldowan flaking requires controlling and coordinating two variables: (1) the force and angle of the hammer stone strike and (2) the orientation of the core (Foucart, Bril, Hirata, Morimura, Houki, Ueno, & Matsuzawa, 2005). Nut cracking requires controlling only the hammer strike. Thus, by 2.6 MBYP, our hominin ancestors had evolved to a level of motor sophistication unique to primates.

Other studies have reinforced this conclusion. Stone tool specialists Nicholas Toth and Kathy Schick compared the motor skills evidenced in bonobo toolmaking with that of Oldowan remains from Gona, Ethiopia (Toth & Schick, 2009). The Gona hominins, they concluded, were significantly better at detaching usable flakes while avoiding miss hits on the core. Oldowan remains from Lokalelei, West Turkana (Kenya) reveal that early toolmakers were highly skilled at adjusting their flaking blows in order to offset minor errors in detachment (Roche, 2005). Nothing comparable has ever been seen in nonhuman ape toolmaking.

Oldowan toolmaking also has an important perceptual dimension. Oldowan knappers did not just grab stones randomly and start whacking away. Instead, they were highly skilled at picking out pebbles that would produce sharp flakes (Schick & Toth, 1993, 126). They could tell whether a stone would make a good core. Accordingly, Oldowan toolmakers were not just acting on their environment in ways unique to primates, they were perceiving it in unique ways as well.

The critical point is that with the emergence of the Oldowan, our ancestors' perceptional/motor repertoire grew larger, sending hominin brain evolution off on a unique trajectory. Functional imaging studies of skilled Oldowan tool manufacture show significant activation in regions of the brain important to object perception and body awareness (the posterior parietal). In addition, those regions associated with manual control and manipulation (the ventral premotor areas) are activated (Stout & Chaminade, 2007; Stout et al., 2008). Expansion of the posterior parietal association cortex, beginning about 3MBYP, may have been an important prerequisite to the emergence of the Oldowan industry (Holloway, Broadfield, & Yuan, 2004). An

important consequence of the Oldowan was to accelerate the evolution of the conscious control of an expanding range of variable motor
movements.

THE HAND AX AND MOTOR CONTROL

Oldowan tools are unquestionably significant in the grand saga of human evolution. But they are not much to look at. You could easily step
over one and never notice. About 1.7 MBYP, a second tool industry, the
Acheulean, emerged that contains artifacts much more clearly identi-
fiable as tools to the modern eye (Lepre et al., 2011). Compared with
the Oldowan, the Acheulean industry emphasized greater concern for
a tool's overall shape, suggesting a steady increase in the control of
motor actions under perceptual/spatial guidance (Wynn, 2002). The
centerpiece of the Acheulean tool kit was the bifacial hand ax, a larger,
much more meticulously fashioned, all-purpose tool. Early hand axes
were rough looking, but by about 500,000 YBP, many of them had become truly stunning, their size, symmetry, and overall aesthetic quality showing evidence of considerable care and skill in their creation
(Foley & Lahr, 2003; Klein, 2005).

An Oldowan flake could be knocked off in a minute or two by a
skilled knapper. Not so with a hand ax. A good hand ax was not easy
to make. Stone tool expert Kathy Schick enumerated the technical
challenges facing the novice ax maker, including properly thinning the
core without removing too much width too soon and controlling the
force of the flaking blows so as to avoid breaking off the tip or splitting
the biface in two (Schick, 1994, 584). Furthermore, some procedural
aspects of hand ax construction, such as thinning the edge, are not
intuitively obvious to beginners (Schick & Toth, 1993, 240). Stephen
Edwards, an experienced stone knapper, claims that many months of
concerted effort are required to reach the skill level of the late Acheu-
leans (Edwards, 2001, 606), often causing beginners to create many
unusable hand axes (Winton, 2005).

Later hand axes often required a sizable investment in time and energy, with toolmakers repeating a series of flaking actions. Wynn (2002) argued that late Acheulean hand ax makers could not simply focus on the tool's edge but needed to understand how flakes trimmed from one part of the stone would affect the tool's overall shape. Thus, they likely had to hold in mind multiple perspectives of the tool as it was being created, constantly adjusting their flaking to meet the changing characteristics of the core as it was shaped (Bril, Roux, & Dietrich, 2005; Pelegrin, 2005). Using refittings, Pelegrin (2009) supports this interpretation, finding evidence that Oldowan tool construction was perceptually controlled and that hand ax construction was conceptually controlled. Along with its technical complexities, the sheer strength and bravery required to make hand axes often are not appreciated. According to Kathy Schick and Nicholas Toth,

> We know from experience that the injuries produced in quarrying massive flakes from boulder cores can be formidable, especially if one is scantily clad. We suspect that death due to loss of blood from a severed artery was probably not unknown in Acheulean times. Accidental injuries from flaking stone may have been one of the most common "occupational hazards" during these times. (1993, 237)

Thus, the technical and physical challenges of constructing a hand ax entailed a significant advance in the cognitive control of motor actions, and neuroscience research supports this conclusion.

Dietrich Stout has been a pioneer in combining neuroscience with archaeology. Using brain-imaging technologies, he revealed what goes on in a person's head as he or she is creating Oldowan and Acheulean tools (Faisal, Stout, Apel, & Bradley, 2010; Stout et al., 2008). His results show that hand ax construction, in contrast to Oldowan flaking, involves significant activation in two areas of the brain necessary for hierarchically organizing and executing motor procedures (the left prefrontal cortex and BA 45 [Broca's analogue] of the right frontal lobe). Furthermore, his studies show that activation in these and other "higher" brain regions cannot be attributed to the "low-level" manipulative demands of making hand axes but instead more likely results from increased cognitive control demands. Put simply, making a hand ax requires more sophisticated thinking and planning than does making an Oldowan flake.

Another cognitive aspect of hand axes distinguishes them from Oldowan flakes. In all likelihood, they were the first tools that were actually conceived of as *tools* (Coolidge & Wynn, 2009, 112). Frederick Coolidge and Tom Wynn—the psychologist/archaeologist partners mentioned earlier—noted that hand axes were treated very differently from Oldowan flakes. Hand axes were often carried around, modified, and resharpened for a variety of different uses. By contrast, Oldowan flakes were usually made on the spot, used, and discarded. Oldowan flakes, they contend, were treated in a way that suggests that they were understood simply as an extension of the arm/hand movement system. They were part of a task procedure and had no real existence outside that context. The hand ax, however, was a separate entity, something distinct from both the user and its uses. One could act both *on* and *with* a hand ax. Thus, the bodily movements used to create and modify the hand ax were conceptually distinct from the movements used when employing the hand ax. It therefore implies an ability to categorize movement types (actions used to create the tool versus actions used with the tool) not present with Oldowan flakes. This ability to conceptually segregate gestures is essential to the process of consciously ritualizing actions.

THE HAND AX AND ACTIVE TEACHING

Given the numerous challenges of hand ax construction, an important question arises—could the skill have been acquired without active teaching? Active teaching is a teacher directly intervening in a learner's behavior in an attempt to make the learner's behavior more effective (e.g., "Here—you do it like this;" Boesch, 1991; Caro & Hauser, 1992). Active teaching is distinguishable from simple imitation (or emulation) or from instances in which a teacher simply encourages or discourages certain behaviors. The emergence of active teaching in hominin evolution represents a major expansion in social cognitive abilities. In chapter 4, I pointed out that adults use a particular type of motor signal—motionese—when transmitting behavioral skills to children. Motionese is slowly paced, stylized, and exaggerated motor movements designed to highlight the behavioral procedures important to some skill, such as how to tie your shoes or shoot a basketball. Motionese is tantamount to ritualized behavior. Thus when we ask whether hand ax construction could be learned without active teaching, we are asking about the earliest origins of consciously ritualized gestures.

Active teaching is nearly nonexistent among nonhuman primates; only one or two possible accounts have been reported (Boesch, 1991). In their review of animal teaching, Caro and Hauser (1992) could find no compelling instances in which a teacher drew a learner's attention to the correct behavior (modeled by the teacher) and then allowed him an opportunity to produce the behavior. The fact that nonhuman apes have been able to produce Oldowan-like tools suggests that this skill probably did not require active teaching to be acquired (Toth & Schick, 2009). However, this does not appear to be true for the later Acheulean hand axes.

In a few traditional societies, stone tools are made that are similar to Acheulean hand axes. One such society is the Kim-Yal of New Guinea, which makes stone ax-heads, called *adzes* (Schick & Toth, 1993, 245–251; Stout, 2005). Making adzes is the exclusive domain of adult males. Learning the adze-making craft usually begins at around age twelve, with an apprenticeship of ten years or more being fairly common. Similarly, stone knappers from Khambat (India) usually begin training at age ten or twelve and spend from three to ten years in apprenticeship (Bril, Roux, & Dietrich, 2005; Roux & David, 2005). In both cases, these apprenticeships involve active guidance by a master using both oral and gestural instruction demonstrating specific hand positions, grips, and striking angles and forces. When we combine this with the evidence already reviewed showing that (1) quality hand axes are not easy to make, (2) some aspects of hand ax construction are not intuitive, and (3) novice hand ax makers often produce many unusable implements, we have a strong case for the necessity of active teaching in hand ax construction. Finally, we should note that by around 300,000 YBP, the archaeological record shows regional styles in hand axes (Foley & Lahr, 2003; Schick & Toth, 1993, 282–283; Wynn & Tierson, 1990). This suggests that the skill was being socially transmitted in local groups, in a way similar to that found among traditional stone-knapping societies such as the Kim-Yal (Stout, 2005).

GESTURE WITHOUT PURPOSE

The method of making a hand ax had to be actively taught; it couldn't be passively acquired. Exactly what was being taught? Unlike Oldowan tool manufacture, making hand axes requires learning elemental gestures that are separate from the immediate purpose. To understand what this means, consider two scenarios from our evolutionary past. In the first scenario, a novice is watching a master make an Oldowan flake. What does the novice observe? He sees another hominin take a hammer stone and strike it against a core stone, thereby producing a flake. He then sees the master examine the edge for sharpness and begin cutting into a hide with it. Moral of the story: a certain gesture— the percussive strike of the hammer stone against the core—produces a useful implement (a sharp edge that cuts).

Now consider the second scenario: A novice observes a master making a hand ax. The master takes a large hammer stone (possibly one requiring both hands to wield), strikes it against an even larger rock stationed on the ground, and breaks off a large flake (call it flake 1). The master then takes a different, smaller, hammer stone and begins removing a series of moderate-sized flakes from flake 1, pausing every now and then to examine the flake's overall shape. This process continues until flake 1 has been reduced to a roughly oval shape with a point beginning to emerge at one end (this is called the "roughing-out" stage, in which the general form of the hand ax takes shape). Then the master precisely removes a series of very thin flakes and puts the finishing touches on the tool, making it as smooth and symmetrical as possible. What is critical about this process is that, unlike Oldowan toolmaking, the percussive striking gesture does not lead directly to a functional tool. Instead, each gesture progressively shapes a core. There is no direct connection between the gesture and the instrumental end. Elemental gestures are embedded in a process, and it is the total process that creates the tool.

Something comparable to this can be seen in the nonhuman ape world, although it is simpler and more primitive. Japanese primatologist Tetsuro Matsuzawa (2001) analyzed chimpanzees' use of tools through a tree structure depicting the relationships among the tool user, the tool, and the object being acted on. For example, a chimp inserting a branch into an ant mound to capture ants represents the simplest relationship (called "level 1"), in which the agent (chimp) uses an instrument (branch) to relate to an object (ants/ant mound). In level 1, the tool-use gesture (inserting the branch) leads directly to a desired outcome: ants to eat. Level 1 is the predominant mode of tool use among chimpanzees. Occasionally, however, a more complicated form of tool use has been observed. When cracking nuts, a chimp sometimes places the nut on a rock and then picks up another rock and pounds on the nut. This is an example of level 2 tool use because the agent (chimp) is juggling two relationships, one between the nut and the rock on which it is set (the anvil) and another between the nut and the rock used to open it (hammer). Even rarer is level 3 tool use, in which a third rock (a wedge) might be employed to straighten the anvil if it is slanted. The importance of levels 2 and 3 tool use is that they involve gestures that are not directly tied to opening the nut, setting the nut on the anvil (what Matsuzawa calls the "positioning" phase), and placing the wedge below the anvil.

This demonstrates that chimpanzees have some capacity to chain together movements leading to an instrumental end, in which the earlier movements in the chain do not directly produce the final result. Note, however, that (1) cracking nuts is rare relative to the totality of chimpanzees' tool use and (2) the end to which the gestures are aimed is constantly within visual access. In other words, the nut to be cracked is visibly present throughout the process. By contrast, making hand axes was not rare. Indeed, some sites contain literally hundreds of hand axes. Furthermore, the hand axe itself did not appear until the final stages of the knapping process. Before that, it was just another rock.

The divorce of intentional actions from obvious instrumental ends became even more pronounced with the creation of composite tools beginning around 300,000 YBP (McBrearty & Tryon, 2006). Composite tools are made of separate elements such as a hafted spear made of a shaft, a sharpened point, and an adhesive or binding material used to affix the point to the shaft. The processes of making a composite tool could extend over hours or even days (Wadley, Hodgskiss, & Grant, 2009).

For example, the South African archaeologist Lyn Wadley and her colleagues analyzed in detail the process of composite tool manufacture used by *Homo sapiens* at Sibudu Cave in southern Africa (dating about 70,000 YBP). These toolmakers were highly skilled and knowledgeable about the mechanics, chemistry, and even the pyrotechnics of composite tool manufacture.

To make the adhesives to attach a spear point to the shaft, they often combined red ochre with tree sap. The ochre's iron content helped produce an adhesive less brittle than tree sap alone, thus making the spear point less likely to shatter on use. Other agents such as beeswax or fat also were frequently added to the mixture in order to ensure the "workability" of the adhesive. Furthermore, once the compound was created, it needed to be dried under well-controlled conditions, since boiling or overdrying the compound would weaken it.

A process this complex necessarily involved numerous intermediate steps far removed from the final product. For example, all the stirring and combining actions were both temporally distant from the creation of the spear (by hours or days) and physically distant in that none of them had any direct contact with the spear itself. Collecting the proper type of wood needed for drying the adhesive is another example of a movement pattern both temporally and physically remote from the creation of the spear. Apart from humans, there are few, if any examples of gestures that have become so distantly divorced from instrumental ends.

When gestures are divorced from ends, they become "freed," conceptually and perceptually distinguishable as separate units. These units can then be associated with the end in such a way as to represent it. An example is someone sitting at a sports bar who looks up at the server and makes a "popping the top off a bottle" gesture. The gesture is a distinct perceptual unit that can stand for drinking a beer. Note also that it is only indirectly connected to drinking a beer but that it is probably enough to convey the message to the server. But suppose the server doesn't understand it. The patron might try another, more direct gesture, the "tipping" motion used when drinking from the bottle. To convey a message more effectively, these distinct-but-associated units can be exaggerated, stylized, and formalized in their presentation. In other words, they are ritualized. Initially, these ritualized units probably remain closely associated with the end goal of which they were originally a part. But over time, their meanings can become more remote. For example, how an extended middle finger acquired its current disparaging meaning is matter of debate, but the meaning is there nonetheless.

From Ritualized Action to Ritual: Dancing for Resources

By around 500,000 to 300,000 YBP, the ability to create handsome hand axes and the complicated procedures required to make composite tools indicate that our hominin ancestors very likely had the necessary motor control abilities to ritualize actions consciously. They could isolate certain gestures; stylize, exaggerate, and formalize those gestures; link them to other gestures; and then coordinate their group-wide expression. For example, they could probably do something equivalent to the *haka* ritual, in which the elemental acts of slapping, stepping, and grunting are isolated, stylized, and combined together in a rhythmic group dance.

Even if our ancestors probably could sing and dance together, did they? Why would they? What would they sing and dance about? They undoubtedly sang and danced as a form of celebration. A successful hunt or a child's birth might be enough reason for the group to sing or dance for joy (for an imaginative example, see Mithen, 2006, 195–196). Human rituals, however, are more than just spontaneous expressions of jubilation. They are rituals, and not just ritualized actions, because they have a symbolic meaning. Ritual participants understand that by performing certain actions in a certain way as a group, they are conveying a message. They are enacting or representing a cultural concept or important value.

Anthropologists Camilla Power and Christopher Knight argue that ritualized singing and dancing may have started to acquire more symbolic, ritual significance around the same time as handsome hand axes and composite tools appeared (500,000 to 250,000 YBP; Knight, Power, & Watts, 1995; Power, 1998). Around this time, our ancestors may have been facing a social crisis serious enough to threaten their very existence. Fossil evidence from this period indicates a major increase in brain size (Ruff, Trinkhaus, & Holliday, 1997), which would have meant that hominin females were giving birth to increasingly helpless infants. The burden of extended and extensive infant care would have severely limited the female's ability to forage, thus making the male's provisioning crucial to her survival and the survival of her offspring.

Imagine the plight of a hominin female with an active toddler and another one on the way. She obviously must rely on provisioning from her male mate. What might be the greatest threat to that provisioning? Power and Knight contend it probably would have been the irresistible charms of younger females in the social group. Controlling that threat so as to ensure the fidelity of her male mate may have been the origin of ritual dance. As a hominin girl began to mature, she would be initiated into adult female society by being taught strenuous ritual dances. Dancing with the other females in group would (1) unify her with the other females in the group—making her first and strongest loyalty to them and not to the males—and (2) signal to the males in the group that all the females (including the new initiates) were unified in their demand for mate provisioning. That is, the dance represented and reinforced a social value: fidelity in male mate provisioning.

Power and Knight's hypothesis is speculative and needs to be viewed with caution. But they do draw on two lines of evidence in order to strengthen it: (1) the occurrence of red ochre and other mineral pigments in the archaeological record and (2) the ubiquity of female initiation ceremonies among traditional societies, especially in southern Africa.

In regard to the first line of evidence, red ochre is a mineral pigment that is widely used by traditional hunter-gatherers in rituals and ceremonies, and evidence of red ochre use among our hominin ancestors can be traced back to about 300,000 YBP (Barham, 2002; Clark & Brown, 2001). While red ochre and other mineral pigments can have practical uses (as we saw earlier with the making of adhesives), particular color pigments seem to have been intentionally gathered at many sites even when this required greater effort and offered little utilitarian payback. This suggests ritual rather than practical use.

In regard to the second line of evidence, numerous traditional societies (e.g., !Kung San, /Xam, Hadza) have female initiation ceremonies
with qualities and themes similar to those discussed by Power and
Knight. Often these ceremonies involve rigorous female-only dances,
seclusion, ritual surgeries, and "secret" knowledge transmitted under
coercive conditions in which failure by the initiate could mean ostra-
cism and the loss of marriageability. All this serves to bind the initiate
closely to her female group members. We will probably never know
for certain whether our ancestors danced 250,000 or so years ago to
ensure male provisioning. But we do know that at some point, dancing
became more than just a physical act. It took on symbolic importance.
It became ritual. One of the strengths of Power and Knight's theory is
that it ties this ritual directly to reproductive success.

Ritual Objects

Revisit for a moment our patron in the sports bar. He used two gestures to try to convey his desire for another beer: popping the top off
of an imaginary beer bottle and tilting the bottle to his mouth. If gestures can be segregated from their instrumental ends so that they can
be used to signal or represent those ends, then so can the objects associated with those ends. Suppose now that the patron does something
a bit different—he points at an empty beer bottle sitting on his table.
It is now an object that represents his desire. Earlier we encountered
Coolidge and Wynn's argument that hand axes were different from
Oldowan tools because they were conceptualized as distinct entities,
as tools. Because the hand ax could be conceptualized as distinct from
its use and its user, it could also serve as a representation of something
with which it was associated (similar to the beer bottle being associated with and therefore representative of drinking a beer). Thus, it is
possible that a hominin might have brandished his hand ax to indi-

cate his desire for food—something quite unlikely with an Oldowan flake.

Objects are not just associated with their activities; they also become associated with their users. Suppose our sports bar patron has a particular mug he drinks from. The mug is connected not only to beer drinking but also to him. Perhaps it is his father's mug, handed down from generation to generation. Suppose his father was a war veteran, with certain ideas about life, loyalty, friendship, duty, and the like. The mug itself might serve as a mental cue to these things. Something similar may have happened with hand axes. At many sites, hundreds of them show little if any signs of use (Klein & Edgar, 2002, 107; Kohen & Mithen, 1999; Stringer & Andrews, 2005, 225). Archaeologist John Hoffecker (2011, 62–63) argues that hand axes were the first externalized ideas that hominins created. Ideas never exist in isolation but always have a context, other ideas with which they are associated. Could a hand ax have been something that was preserved and passed down from father to son? Could something of the father's personality and behavior—his "essence"—have been captured in the tool?

In this way, mere objects become *ritual* objects. Even though hand axes might have served ritual purposes, we have no good evidence to support this. Arguably, the first evidence of ritual objects dates to around 160,000 YBP, in the form of three *Homo sapiens* skulls unearthed in Herto, Ethiopia (White, Asfaw, & DeGusta, 2003). These are some of the oldest fossils attributed to modern *Homo sapiens*. What's intriguing about the Herto skulls is that they bear possible evidence of ritualistic behavior. First, at least one of the skulls shows evidence of defleshing; that is, a stone tool was used to deliberately remove the flesh. Defleshing requires some effort and has very little return in terms of acquired nutrition, which leaves open the possibility that it was for ritual, not practical, purposes. Furthermore, the outer surfaces of all three skulls show evidence of being polished, indicating that they were rubbed for a prolonged period against another surface.

This would happen if they were being carried around, in a bag, perhaps. Why carry skulls around in a bag?

One possibility is that the skulls carried some symbolic significance to the possessor. From prehistoric times to the present, relics have been prized human possessions. Relics are objects or body parts associated with some important individual, place, or event. Relics from individuals are often thought to possess some essence of the individual. In many religions, the relic of a saint or holy person is believed to have a special spiritual value. It is somehow holy, connected to the divine in a way that ordinary objects are not. In ancient Greece, the armor or weapons of a great warrior were often treated as relics. So, too, were objects thought to be sacred to a patron deity, often believed to be critical to a city's protection and prosperity. The deity's relic was often displayed in a holy place (temple or sanctuary) and brought out on ceremonial occasions.

During the Upper Paleolithic period (35,000 to 10,000 YBP), a variety of what appear to be purely ornamental or ceremonial artifacts emerge in the archaeological record. Carved antler batons; half-human / half-animal carvings, drawings, and figurines; and finely shaped blades too thin to be of practical use are among what appear to be ritual objects (White, 2003). These objects probably served a variety of purposes, such as signaling rank, representing gods or spirits, and celebrating power or fertility. Another ritual purpose, however, can be detected at the Upper Paleolithic site of Saint-Germain-la-Rivière (France). Here a female was buried with scores of finely crafted and perforated deer-tooth ornaments (Vanhaeren & d'Errico, 2005). What's significant about the ornaments is that their origin is hundreds of miles from the burial site. They are "exotic" imports, brought to the site via extended trade networks.

Beginning about 70,000 YBP and extending through the Upper Paleolithic, we have the beginnings of intergroup trade (Ambrose, 2002). Practical items were undoubtedly among the objects traded, but not always. Sometimes the traded objects were not practical but ceremonial or ornamental, such as finely made tools, beads, and body ornaments like the deer-tooth pendants of Saint-Germain-la-Rivière (Vanhaeren et al., 2006). Why trade objects with little practical use? The answer is that trade serves a ritual as well as a practical purpose. When President Richard Nixon visited China, why did the Chinese give a pair of pandas to the Americans? And why did the Americans reciprocate with a pair of musk ox? Hardly practical items! Trading gift items—items whose value is in their appearance, rarity, or craftsmanship—carries symbolic meaning. It's ritual trade, designed to build trust. Before there can be commercial trade, there must be trust. Establishing relations with an out-group represents another socially stressful transition, and ritual trade helps ensure that the transition goes smoothly.

SCENE 8: CAN ROBERTO BE TRUSTED?

By nature, Roberto Canessa was intense and volatile, qualities that made him a force on the rugby field but trying and unpredictable off it. A few days into their final expedition across the Andes, as Parrado and Canessa's strength and hope faded, Canessa pointed to the belt he was wearing. It was Panchito's. He was your best friend, Canessa told Parrado. Now I am your best friend. (as told in *Alive*, 54 and 238; and in *Miracle in the Andes*, 267–268)

Perhaps no one had as widespread an impact on the group as Roberto Canessa did. Within the mountain social system, Canessa would not allow himself to be limited to any particular role or responsibility. Medical authority, inventor, expeditionary, amateur theologian, and moral philosopher—wherever he deemed his skills were needed, Roberto was there—invited or not. This was a heavy (often self-imposed) burden for a nineteen-year-old, but his physical and mental assets made Roberto a potentially pivotal character in the group's success.

But if unfocused or wrongly directed, his immense energies could be dangerous. Could Roberto be trusted? Could the group put its fragile hopes in his hands?

No one felt these concerns more acutely than Nando Parrado. He needed a partner to climb the mountain, and he needed Roberto. But before the crash, he and Roberto were not the closest of friends, and their expedition would be a suicidal disaster if Roberto's passions raged uncontrollably or were wasted on petty squabbling. Parrado needed him to direct all his wits and fire at the mountain and at death. If he could somehow get Roberto do that, then together they just might have a fighting chance. Roberto was too smart not to know all this. He had to know how much others were counting on him and that his volatility worried them. He knew the stakes they were playing for, that all their lives—Parrado's especially—were in the balance. How could he assure them? How could he convince Parrado and the others that he wouldn't let them down? How could he tell them that his love for them would not be overpowered by his own hot-blooded nature? Ritual again was the answer. Throughout human history, when words were inadequate to give assurance, ritual could satisfy. Accordingly, Roberto would demonstrate his commitment with ritual.

A personal relic gains its ritual status because of its association with an individual. The bones of saints, splinters of the true cross, Muhammad's cloak all are regarded as personal relics endowed with spiritual significance for believers. The power of the relic is in the belief that it carries the essence of the individual. Chapter 2 explained that to some extent, everyone is susceptible to magical contagion, believing that objects or articles of clothing associated with a certain individual possess and transmit something of the character of that individual.

In book XVI of the *Iliad*, Patroklos dons Achilles's armor in order to fend off the Trojans, who have drawn perilously close to the Greek's ships. As he does so, Patroklos merely says that he wishes to deceive the Trojans into thinking that the great Achilles had returned to battle.

Classics scholars, however, tell us that there is far more to the donning of Achilles armor than simply a ruse (Paton, 1912).

The armor is enchanted, having been forged by the god Hephaestus. Likewise, Achilles himself is something more than just a mortal, being the son of a goddess. By wearing Achilles's armor, Patroklos not only benefits from its divine power, but he also becomes Achilles-like in his fighting prowess. He cannot be killed until the armor is stripped from him (by the god Apollo), just as Achilles cannot be killed without divine assistance. The armor carries something of Achilles's essence, something of his semidivine nature. But the armor also carries with it responsibility. Achilles instructs Patroklos not to pursue the Trojans back to the gates of their city, but to push them back only far enough to save the ships, an admonition that in the heat of battle Patroklos chooses to ignore. For that lapse in judgment, he pays with his life. As an extension of Achilles, the armor both benefits and burdens its bearer—it bestows an essence of Achilles, but it also imposes a responsibility. Such is the dual nature of ritual objects. The royal crown is power but also the weight of responsibility.

We are given no details about the circumstances under which Roberto Canessa took Panchito Abal's belt.[2] Canessa had lost nearly forty pounds on the mountain, and he may have needed it merely to keep his pants up. Then again, maybe the act was more deliberate. In any case, only a few days into their journey, Roberto recognized the belt's significance. It wasn't magical. It wasn't Achilles's armor. But it did bring with it a certain responsibility by representing a transfer of roles. A new sacred commitment had been forged. Canessa could have said nothing about the belt. He could have just walked on behind Parrado that day as he had before. But he purposely called attention to the belt's significance, and with that, he acquired some of Panchito's spirit and assumed Panchito's role.

On the ninth day of their journey, Canessa collapsed, overcome by exhaustion and diarrhea. Although Parrado took his pack, he made it clear that he had to push on; to stop now would probably mean the end for both of them. Where do we find the strength to go on when the body is broken? Parrado marched on but kept an eye on his best friend. Somehow Canessa made it back to his feet and forced himself forward, Parrado tells us, through nothing but "stubbornness and the sheer power of his will" (219). He couldn't let his best friend down.

CHAPTER SIX

Ritual Defeats the Mountain

SCENE 9: NEVER MIND THE PAIN

Two weary figures plod along in the snow, one about twenty paces in front of the other. Their breathing is heavy and labored but also steady and rhythmic. Listening carefully as they pass, one can detect audible words spoken in unison with each slow, effortful step. "Hail Mary, full of grace . . ." groans the first as he passes; ". . . on earth as it is in heaven . . ." the second says as his foot crushes into the snow. Suddenly, the trailing figure falls and a fearful silence ensues. The leader stops and motions to his companion to get up. For a moment, they just stare at each other. "Is this the end?" they both wonder. Slowly the second figure struggles to his feet and takes a step. "Give us this day . . ." he continues, and they walk on. (as told in *Alive*, 281; and *Miracle in the Andes*, 211–212)

On the morning of December 12, 1972, nearly two months to the day since the crash, Nando Parrado, Roberto Canessa, and Antonio Vizintin departed from the wreckage of the Fairchild. Their objective was to climb up and over the 15,000-plus-foot mountain peak immediately west of the Fairchild and descend into what they expected to be the green valleys of Chile. They would then summon rescuers to retrieve the remaining thirteen of their companions. This was the eighth expedition to set out across the Andes. It was the last only because it succeeded. Had it failed, they would have kept trying. But everyone understood that this expedition was their best chance. Although unspoken, they also understood that it was a very slim chance.

The mountain they were to climb had not been named. Parrado, the first to reach the summit, named it Mount Seler for his father. Although it was not Everest, it was a serious mountain peak for which experienced climbers would have trained, planned, and thoroughly equipped themselves. Seasoned mountaineers know that even when all the proper precautions are taken, climbing can still be fatal. Not Parrado, Canessa, or Vizintin had ever climbed a mountain before; in fact, Canessa and Vizintin had never even touched snow before the crash. They had prepared physically as best as they could, but none of them was in the kind of physical shape necessary to climb a mountain. Finally, needless to say, they had no real climbing equipment. All they had was cobbled together from the remnants of the Fairchild. They were little more than three teenagers in tennis shoes setting out to cross the Andes. Any climber with an ounce of sense would have seen this as suicidal idiocy. But for sixteen boys trapped for two months at eleven thousand feet, there was little choice. Only one thing was certain about their expedition; they were going to suffer immeasurably.

As the boys ascended, the thin cold air made them shiver uncontrollably, and slogging through the deep snow quickly led to exhaustion. They could go only two or three steps before having to stop to catch their breath. Their fingers quickly went numb from the cold, and their limbs grew heavy and clumsy. Parrado remembered that simply turning his head to speak to Roberto left him gasping for breath, and no matter how deeply he inhaled, his lungs never seemed to fill. The high altitude accelerated their heart rate, thickened their blood, and brought them close to hyperventilation. In addition, the dry, rarified air quickly dehydrated them. Well-prepared mountain climbers would have brought portable gas stoves to melt snow for drinking water. But the boys could only gulp handfuls of snow as they went, which did not even come close to satisfying their persistent thirst.

At night, to keep from freezing, the three crammed into a single patched-together sleeping bag. The tight squeeze was achingly uncomfortable but necessary to stave off hypothermia. "We kept ourselves from freezing, but still we suffered terribly," Parrado tells us (*Miracle in the Andes*, 195). Then, of course, there was the constant walking—climbing up the mountain, trudging down, and endlessly pacing across the long valley at the mountain's base. Sore feet, cramped legs, and chapped, bleeding skin caused pain with every weary step in what turned out to be a ten-day march across the frozen, rocky Andes landscape. A march on which, Parrado noted, they were simply walking themselves to death. But they were not without weapons in this war of attrition. They had an ally—their minds. Ritual can harness the mind's power to endure. Parrado and Canessa discovered this along the way. Tibetan monks, however, have known it for centuries.

The Ritual Training of Attention

In the mid-1980s, Herbert Benson, an associate professor of medicine at Harvard Medical School, traveled to the Himalayas to study Tibetan monks' meditation practices. What he found astounded him. On a rocky ledge some fifteen thousand feet above sea level, monks, dressed only in woolen or cotton shawls, were "camping out" in overnight temperatures that dipped to near zero degrees Fahrenheit.[1] The monks did not huddle together, shiver, or show any signs of discomfort associated with the cold. Benson also documented how the monks, meditating in chilly monastic chambers of only 40 degrees Fahrenheit, used their body heat to dry cold, soaking-wet towels placed on their backs. The monks practiced a form of yogic meditation called Tummo, during which their body temperature, measured at the fingers and toes, increased by as much as seventeen degrees. Drying towels and sleeping nearly naked in the freezing cold were not parlor tricks but demonstrations of the mind's power to access realities existing beyond human senses and emotions. In everyday reality, our attention is preoccupied by sensory stimulation and emotional desire, which blinds us to the true extent of the mind's power. By training themselves to focus their attention inward, the monks were able to access the mind's buried energy and use it to generate heat.

At the heart of any ritual is focusing one's attention. I worked at a car wash when I was in high school. I could wash windows "automatically": Spray on the cleaner, wipe until the window is clear, move on to the next one, repeat. What guides the action is the end result. You just keep your hand moving until your eyes tell you that the window is clear and then move on—no thinking required. Ritual washing is not window washing. By making it into a ritual, the attention is focused on the act itself. The act is the end. By washing in a specific way (three strokes up, three down), ritual washing signals respect for the object, not the importance of cleanliness.

Take another example: the three-shot rifle volley at military funerals (often mistakenly called a "twenty-one gun salute"). To shoot a rifle, you have to bring it into firing position. It's the end result—getting the butt against your shoulder and the sight at eye level—that tells you that you have done it right. But if you are part of the honor guard at a military funeral, simply getting the rifle into a firing position is not necessarily "doing it right." Instead, there's a very strict protocol about how the rifle is brought into firing position, fired, reloaded, and fired again. The ritualized nature of the act is an end unto itself. It signals respect for the dead, just as ritual washing signals reverence for the object. Ritual is effective in sending these messages because it takes what are normally automatic, thoughtless, end-guided actions and forces attention on their execution, thereby making them deliberate, controlled, thoughtful, and meaningful.

By forcing attention on the details of action, ritual can "fill" our conscious awareness, preventing us from attending to extraneous, irrelevant, or threatening signals or thoughts. This is one of the reasons why athletes use ritual so often. If you keep your mind focused on the details of the proper action, then the proper end result usually will follow. Old basketball fans may remember Adrian Dantley, a star forward with Notre Dame in the 1970s and later with the Utah Jazz (and others) in the NBA. Dantley had a strict foul-line ritual: bounce the ball three times, bring it to the chest while looking at the rim, recite "over the front of the rim, backspin, follow through," and then shoot. This ritual forced Dantley to focus on the specifics of the free throw while at the same time directing his attention away from hooting fans, negative thoughts, or emotions associated with success or failure. It worked. He shot more than 80 percent from the foul line in his NBA career. Because ritual requires the focusing of attention, consistently practiced ritual strengthens and enhances this ability.

The ritualized practices associated with meditation are very likely the oldest techniques for training attention. There are many forms of meditation, but in general, meditative practices can be categorized as either *concentration* or *mindfulness* (Barendregt, 2011; Lutz, Slagter, Dunne, & Davidson, 2008). In concentration meditation, you train your attention by deliberately focusing on some object, image, process (such as breathing), or phrase (a mantra or prayer). Mindfulness meditation involves cultivating a nonjudgmental awareness of the present moment or what might be called an "open" monitoring of awareness. Although the differences between them are not trivial, concentration and mindfulness meditation also overlap in many respects. Both train and enhance attentional control. Mindfulness meditation often uses focusing of attention on a single point (such as the breathing process) to begin developing present-moment nonjudgmental awareness.

In both concentration and mindfulness meditation, a common starting point is the simple monitoring of breathing. The most "low level," elemental, and automatic of bodily processes is brought into conscious awareness. Learning to focus attention on the breathing process and experiencing deeply its varied aspects, like the feeling of the air at the nostrils, the movement of the diaphragm, and the filling of the lungs, are often the first goals of meditation. Frequently, the breathing process is connected to positive thoughts or ideals, so one breathes in peace and exhales light. Or the breaths are connected rhythmically to a prayer or mantra. If one's attention is momentarily taken away from the breathing process—a barking dog outside or an involuntary "naughty" thought—one simply notes the disruption without becoming emotionally involved with it and gently returns to the monitoring of breathing. This may seem quite simple, but such simple beginnings can lead to the extraordinary feats of attentional control shown by the Tibetan monks. In Tummo meditation, for example, monks learn to clear their minds entirely while harnessing and directing their mental energy toward the production of body heat. Part of this process is the simple monitoring of breathing, in which the mind is fed with energy from the breath in order to generate heat.

Attentional Expertise

Research has found that advanced meditators are able to engage in not only astounding feats of physical endurance and pain control but great attentional feats as well. For example, when two different images are projected onto the retina of each eye, it causes binocular rivalry. Normally our eyes work together, fixating on a single object or event in the environment. Thus, the same image is projected onto the retina of each eye. But when different images are projected onto each retina, the visual system goes into conflict with itself, that is, binocular rivalry. In binocular rivalry, our awareness of the two different images alternates as they fight for limited attentional resources. So if one eye sees the letter "O" and the other "X," our awareness will constantly shift between O and X. It's an involuntary switching of attention. We can't control which pattern is seen; they just continually shift over time. At least, that is what we always believed. Experienced concentration meditators, however, are able to "choose" which image they perceive during binocular rivalry (Carter et al., 2005). In other words, they have trained their attentional system to such an extent that they can exert a certain degree of voluntary control over what is normally an involuntary attentional switching.

Attentional blink, another "involuntary" attentional phenomenon, is the degraded ability or inability to perceive the second of two sequentially presented visual patterns. For example, suppose I flash a picture of Donald Duck at you, followed a half second later by a picture of Bugs Bunny. Because Bugs follows so quickly after Donald, often you will not detect it at all or will detect it only partially (hence attentional blink—blink and you'll miss it; Martens, Munneke, Smid, & Johnson, 2006; Shapiro, Arnell, & Raymond, 1997). The explanation for attentional blink is that attention has limited resources (you can pay attention to only so many things), so when a second visual pattern follows the first one too quickly, all your resources will still be busy processing the first. As with binocular rivalry, attentional blink was generally thought to be involuntary. However, three months of intense (ten to twelve hours a day) mindfulness meditation training was found to significantly reduce attentional blink (Slagter et al., 2007). Meditators more accurately identified the second target while suffering no decrement in performance on the first. By monitoring brain waves, the researchers found that those meditators who performed the best were those most able to direct their attentional resources away from the first to the second pattern.

Meditation has also been found to improve *attentional vigilance,* the ability to sustain attention over time and to respond appropriately to randomly presented signals (Grier et al., 2003; Mackworth, 1948). Typically, as a vigilance task wears on, the detection of signals declines and reaction times increase. That is, the longer you perform a tedious task, the more likely your mind will start to wander, causing you to miss signals and/or react more slowly to them. Remaining vigilant is even harder if you are sleep deprived. Real-world examples of vigilance tasks are monitoring a radar screen or a security camera. Even though the jobs are important and the signals critical, the overall task is monotonous.

University of Kentucky biologist Bruce O'Hara wondered if meditation could improve performance on vigilance tasks and/or reverse sleep-related declines in performance (Kaul, Passafiume, Sargent, & O'Hara, 2010). Accordingly, he had his subjects perform a vigilance task after a nap, concentration meditation, or a control activity (sitting comfortably with light music on while either reading or conversing). Significantly better performance was found in the meditation condition compared with either the nap or the control conditions. Furthermore, meditators suffered significantly less performance decline after losing a night's sleep.

Finally, neuroscience evidence directly links meditation with attentional control. Researchers at the University of Wisconsin (Brefczynski-Lewis, Lutz, Schaefer, Levinson, & Davidson, 2007) monitored brain activity in highly practiced meditators (with an average of 44,000 hours of meditation), modestly practiced meditators (an average of 19,000 hours), and novice meditators (only a week's practice before the study) while they engaged in concentration meditation. All the subjects showed increased activity in a broad network of brain circuits associated with attentional control (frontal/parietal regions, lateral occipital, insula, thalamus, basal ganglia, and cerebellum). The degree of activation of these areas followed a general inverted U-shaped function. That is, the modestly practiced meditators showed significantly more activation in the attention network than did the novices and the highly practiced meditators. In fact, the highly practiced meditators had the least activation of all, and they also showed the least emotional reaction to distracting sounds. These results indicate that (1) as expected, meditation activates attentional systems and (2) as one gains more experience in meditation, the efficiency of the attentional system increases so that less effort and energy at the brain level are required to exert attentional control, thereby decreasing activation in the attention network of highly practiced meditators. Highly practiced meditators become "attentional experts."

Attentional expertise is behind the monks' ability to endure what for the rest of us would be intolerable discomfort. They literally *do not mind the pain.* In other words, they don't pay attention to it. By evolutionary design, pain signals have an involuntary pull on attention; we can't help minding the pain. This is good because pain is usually a signal of tissue damage. Pain tells us that we need to take some action in order to reduce bodily harm that could threaten our survival. But chronic, nonlethal pain can be a living hell, and a number of studies have shown how the redirection of attention, through meditation, can be an effective way of relieving pain.

Dr. John Kabat-Zinn at the University of Massachusetts Hospital (Kabat-Zinn, 1982) assigned fifty-one chronic pain patients to a ten-week mindfulness meditation program. The patients had a variety of complaints, ranging from lower back, neck and shoulder, facial, noncoronary chest, and GI tract pain, none of which had responded favorably to traditional medical treatments. At the end of the ten weeks, 65 percent of the patients had at least a one-third measured reduction in experienced pain. A more recent study monitored brain activity while subjects engaged in mindfulness meditation as they were intermittently exposed to an unpleasant heat stimulus (Zeidan et al., 2011). Meditation was found to significantly reduce both pain and the unpleasantness ratings of the heat stimulus. Furthermore, these reductions were associated with less activation in brain areas associated with sensory and emotional processing (the somato-senory cortex and thalamus) and more activation in areas associated with the cognitive interpretation of sensory signals (orbito-frontal cortex) and pain regulation (insula and anterior cingulate cortex). Put more simply, meditation trained the brain to attend to the pain more as an "observer" and less as an "experience-er."

A Long, Grueling Ritual

Parrado and Canessa took three days to reach the summit of Mount Seler. Their success unleashed a torrent of mixed emotions. They were literally standing on top (of this particular part) of the world, but their sense of awe faded quickly as they realized that there were no green valleys of Chile below them, no sign of life or civilization anywhere. They cursed God and fell to their knees in despair. But as they stared across the vast expanse of the Andes, they made out a valley snaking its way through the mountains. Far into the distance, the snow-capped peaks lining the valley gave way to smaller, rocky gray ones. The valley was the way out. But it was a long, long way off, and they hadn't planned for such a protracted journey. They would need more food, more supplies. There wouldn't be enough for all three of them to make it all the way, if, in fact, they could make it all the way. The next morning, Parrado and Canessa took Vizintin's supplies and sent him back to the plane. Then the two commenced their arduous trek down the mountain and into the valley. How do you make such an impossible journey? Using strength, faith, determination, courage, and even desperation. But it was on this incredible journey that Parrado and Canessa used another uniquely human survival weapon: ritual. The journey was a long, grueling ritual of mind over death.

Stay in the Moment

Ritual forces one to focus on the present, on the process, on the elemental gestures of an action sequence. In ritual, the action is not a means to an end, it is the end.

This emphasis on the present prevents us from attending to extraneous thoughts that could be distracting or debilitating. Future thinking is pushed out of awareness, and only the present matters. Patients with chronic pain who use mindfulness coping strategies are taught to attend only to the "now" of their pain experience. Future thinking magnifies pain, and we grow despondent thinking that the pain will "never end" or that it will be impossible to endure the pain "all night." But we need not endure the pain "all night," only for "this moment."

The climb up Mount Seler was laborious; hours of painful, exhausting climbing and little progress. The mountain's vast scale fooled them into misjudging distances, so something that seemed only yards away was really a mile or more ahead. False peaks would make them think they were near the top but then crush their spirits with more mountain to climb. Repeatedly, the expeditionaries were seized with panic and hopelessness at the impossibility of their task. Future thinking was destroying them. "You're drowning in distance," Parrado would say to himself. To cope, he forced himself to stay in the present. He reduced the mountain to the current step, then the next step, and then the rock just ahead. What he began to realize was *no one can climb a mountain*. Rather, one climbs to the next landmark, step by step.

Attending to the Elemental Gestures

Ritual forces attention on the "lowest level" gestures. You concentrate on the details of breathing or the process of bringing the rifle to your shoulder. The gestures don't achieve a goal; they are the goal. Parrado tells us that as he walked, his entire consciousness narrowed to nothing but his breathing and the rhythmic crunch of his feet in the snow. To this he would sometimes add a mantra—a constant, steady repetition of Hail Marys (for Canessa it was the Lord's Prayer). As often happens when intensely applied ritual practices are combined with physical and mental stress, Parrado went into an altered state of consciousness—a trancelike state, he tells us (*Miracle in the Andes*, 190, 212). In that state, he became an observer of his suffering. His fatigue, his longing for his father, and his sense of hopelessness became "like a voice on the radio, playing in another room" (190). He could see images of himself suffering and dying, and yet they aroused no distress. He was detached from them. Instead, his mind was consumed in the repetition of step, breath, and chant. Parrado may have achieved what some meditators refer to as "walking meditation," in which practitioners seek to achieve a state of detached awareness of bodily sensations, thoughts, and feelings as they walk.

Attentional Redirection

Concentration meditation often starts by focusing attention on breathing. The meditator makes himself aware of the incoming breath, the sensation at the nostrils, the filling of the lungs, and other normally unconscious aspects of breathing. Inevitably (especially with novice meditators), there will be shifts in attention. Often without realizing it, the meditator begins thinking about dinner, an ache in the back, or yesterday's softball game. When this occurs, the meditator is instructed to take note of the attentional shift without emotionally engaging with it and without chastising himself for the attentional lapse. Instead, he simply acknowledges the intruding thought and gently returns to monitoring his breathing. With practice, he becomes increasingly proficient at both maintaining a singular attentional focus (on his breathing, a mantra, an image, etc.) and dispassionately disengaging and redirecting attention when the focus shifts. In this way, skill at attentional control develops until eventually it becomes almost effortless.

As the endless march wore on, Canessa's fatigue became overwhelming. More than once he pleaded with Parrado to stop. We must rest, he would say, or we both will collapse soon. Sometimes Parrado would concede and they would pause. Other times, however, like a man possessed, he would insist that they keep moving, whatever the cost. Take your mind off the walking, he would instruct; think of something else. Canessa's attentional redirection took a couple of different forms. He focused on an image—imagining himself walking down the streets of Montevideo, passing the shops, gazing in the windows at the merchandise. At times, his immersion in the image seemed nearly complete. Annoyed at his slow pace, Parrado would call to him to hurry up. No, I can't, Canessa would shout back, I'll miss something in the shops. Other times, he engaged in a continuous dialogue with God, a strategy he picked up from the character Tevye in *Fiddler on the Roof*. Yes, yes, he would tell the Almighty, test us to the very limits of our endurance. But don't make it impossible.

Like Tevye, Canessa addressed God as a friend. God is different from other friends, however, in that he is supernatural. Is this difference important? As a coping strategy, would Canessa's dialogue have been less effective if he had simply imagined a conversation with his father or with a human friend? There are reasons to suspect that it might have been.

Meditative practices are known to have positive health benefits such as lower blood pressure, a reduced heart rate and better mental health. A meta-analysis comparing spiritual (though not necessarily religious) meditation and secular meditation found that these effects tended to occur more frequently with spiritual meditation (Alexander, Rainforth, & Gelderloos, 1991). Another study compared the effects of devotional prayer with those of progressive relaxation training on anxiety, anger, and muscle tension in Christian subjects (Carlson,

Bacaseta, & Simanton, 1988). At the end of a two-week period, those who engaged in prayer had significantly lower anxiety, anger, and muscle tension.

A more direct test of the role of spiritualism was undertaken by psychologists Amy Wachholtz and Ken Pargament (Wachholtz & Pargament, 2005), who compared the effects of spiritual and secular meditation on anxiety reduction, pain tolerance, and mood. Volunteers were randomly assigned to one of three groups practicing different meditative or relaxation techniques. One group concentrated on a spiritual phrase such as "God is love" or "God is peace," while another used a more secular mantra like "I am happy" or "I am joyful." The third group—the control group—was simply given instructions on how to relax. After practicing their technique for twenty minutes a day for two weeks, the subjects were tested on their anxiety, mood, and pain tolerance. Pain tolerance was measured by the amount of time that volunteers could keep their hands in water at a temperature of 35.6 degrees Fahrenheit. Those who had been practicing spiritual meditation were able to keep their hands in the near-freezing water almost twice as long as those in the other groups. In addition, the spiritual group showed greater anxiety reduction and mood elevation.

Wacholtz and Pargament readily acknowledged that one drawback of their study is that their spiritual meditation had relational elements that the secular meditation did not; that is, the spiritual meditators invoked "God," who, for most, was an agent to whom they could relate. Nothing similar exists in secular practices, so the secular meditation condition was more self-centered ("I am joyful"). This methodological weakness, however, may be unavoidable. The presence of supernatural agents to whom one can relate is the essence of what makes (most) religious and spiritual practices "religious" and/or "spiritual."

Social interactions are among the most mentally and attentionally demanding activities in which humans engage. If Canessa had not believed that he was having a "real" social interaction, it may not have worked as a strategy for taking his mind off his torment. Indeed, at the brain level, conversations with supernatural agents are as real as any we have with friends or colleagues, and they are different from those with purely imaginative characters like Santa Claus (Kapogiannis et al., 2009; Schjoedt, Stodkilde-Jorgenson, Geertz, & Reopstorff, 2009).

An important lesson relevant to the evolution of attention emerges from Canessa's divine dialogue, that social factors are ultimately the most responsible for the attentional capacities of *Homo sapiens*. Foraging, avoiding predation, and constructing tools all played a role in the evolution of hominins' attentional control, but in the end, it was the social world that made us what we are today.

Evolution of Attentional Control

Attention is the ability to direct mental resources on a particular signal or process. It's what you're doing as you read this text or what a cat does when it stalks a mouse. Attention is selective. As you read this text, you filter out other signals, so you're not thinking about last night's softball game or listening to the advertisement on the radio. At least three distinct brain systems and processes are involved in attention (Dukas, 2009): (1) Alerting: achieving an active mental state in which attention can be directed and sustained. Fatigue, boredom, or drugs can interfere with one's ability to sustain or direct attention. (2) Orienting: focusing mental energy on a selected signal or process while filtering out competing or irrelevant signals or processes. This is what we typically think of when we think of attention: focusing and filtering, concentrating on this and not that. (3) Executive control: resolving conflicts among competing signals or processes. Even after attention is focused on one thing, other signals and processes must continue to be monitored at a low level in case attentional switching is necessary. So even when a mother is talking to her friend on the phone, she keeps an "ear out" for sounds of distress or waking from the nursery. If she hears such a signal, she must then decide whether it warrants ending the conversation.

Predators and Prey

The proper allocation of attentional resources is a widespread challenge across the animal world. For example, as a bird pokes around the ground for worms or bugs to eat, it must also be on guard not to get eaten itself. Thus, it faces an attentional allocation problem: Attending exclusively to foraging could make it dangerously vulnerable to predation, but attending too closely to predation concerns could mean that it starves to death. Successful foragers must find a way of allocating attention so as to maximize resource intake, which is called the "attentive prey" model of attention (Dukas & Ellner, 1993; Stephens & Krebs, 1986). This model can be set up as a simple mathematical formula:

energy acquired from the prey − energy resources necessary to detect and acquire prey = net energy acquired.

The animal's task is to maximize net energy acquired.

The model predicts that as prey camouflage themselves more effectively, predators must dedicate more attentional resources to detecting them. In other words, a predator who feeds on cryptic prey (something well camouflaged and hard to detect) will have to focus its attention more effectively than will a predator that feeds on conspicuous, easy-to-detect prey. Studies support this notion. In the presence of numerous conspicuous prey, predators spread their attention broadly, and in the presence of a single cryptic (but highly valued) prey, predators focus their attention narrowly (Dukas & Ellner, 1993). Thus, we would expect predators that feed on cryptic or otherwise mentally challenging prey to have a greater capacity for focusing and sustaining attention on a single object or event.

Evidence also shows that when more demands are made on attention, more attentional resources are required. In one study, blue jays were given the primary task of trying to detect a highly valued prey (a caterpillar) that was either conspicuous or cryptic (Dukas & Kamil, 2000, 159). In addition, they were simultaneously required to detect a cryptic moth. The results showed that if the caterpillar was conspicuous, the jays could maintain a high detection rate for both it and the cryptic moth. But if the caterpillar was cryptic, they had to reduce their detection rate for the moth in order to maintain a high detection rate for the caterpillar. Simply put, two demanding tasks require more attentional resources than do one easy and one demanding task.

These research findings allow us to ask, Under what circumstances would we expect the greatest selection pressure on attentional capacity? This answer is twofold: (1) We would expect selection pressure on attentional resources to be the greatest for those creatures that must regularly balance two important attentional tasks. One fairly common example of this is creatures that are both predators and prey. Every day, their survival requires striking the right balance between getting lunch while not becoming lunch. (2) We would expect more selection pressure when these creatures' prey are cryptic. Our early ancestors were creatures caught in exactly these circumstances.

About four million years ago, our ancestors were smallish, bipedal apes called *australopiths*. The early australopiths, such as *Australopithecus anamensis* and *Australopithecus afarensis*, were vegetarians, subsisting primarily on fruits, berries, and other plant matter. Later species had more varied diets. *Australopithecus robustus*, which emerged around two million years ago, also ate roots, tubers, and more meat, probably in the form of termites and other insects (Stringer & Andrews, 2005, 127–128).

Tubers, roots, and other underground storage organs (USOs) are significant as food sources because they must be dug up from the ground, either by hand or with tools like digging sticks (Wrangham, Jones, Laden, Pilbeam, & Conklin-Brittain, 1999). Indeed, some evidence shows that later australopiths may have used digging sticks to obtain resources (Brain, 1988). Similarly, termites and other bugs often must be extracted from mounds or from beneath tree bark. Thus, USOs, termites, and other bugs are cryptic prey requiring more attentional effort than does simply picking fruit or berries. But while later australopiths were scouring the ground and trees for cryptic prey, they themselves were the targets of predators (Stringer & Andrews, 2005, 125; Klein & Edgar, 2002, 41). Dramatic evidence of this can be found on the skull of an immature *A. africanus* (Raymond Dart's famous Taung child), which sustained damage consistent with an attack by a predatory bird. For the later australopiths, the proper allocation of attentional resources clearly would have been critical to survival. While foraging, they would have had to devote a large proportion of their attentional focus on cryptic prey while at the same time watching out for larger predators stalking them.

Tools and Attention

Australopithecus robustus may also have been the first to construct stone tools (see Klein & Edgar, 2002, 76–80). As discussed in chapter 5, the earliest stone tools probably do not represent a substantial leap in cognitive abilities over those of extant nonhuman primates. But one aspect of stone tool construction is relevant to the evolution of attentional capacity: the selection of cores. Oldowan toolmakers were highly adept at selecting cores with the right physical properties for producing sharp flakes. Often these cores were found near water sources frequented by large predators. This was another circumstance in which attentional resources had to be allocated judiciously, because the predators always were waiting to pick off unwary prey.

Given the length of time necessary to create a hand ax relative to knocking off an Oldowan flake, attentional capacity very likely expanded with the emergence of the Acheulean industry. Around 400,000 years ago, in the later stages of the Acheulean, that some intriguing archaeological remains point to a significant leap in attentional capacities. At a place called Boxgrove in England, a hominin—a member of the species *Homo heidelbergensis*—crouched down and began removing flakes from a core stone. What remains is a "debitage pile," the discarded flakes resulting from the process of shaping a roughed-out core into a finished hand ax (Coolidge & Wynn, 2009, 159–160). What also is clear, and quite captivating, is that larger, well-shaped flakes are absent from the pile. Evidently, the toolmaker shaped the core but at the same time kept an eye out for well-shaped flakes that were themselves useful. He apparently set aside these nicely shaped flakes and took them with him when he was finished.

What does all this have to do with attention? The answer is that the hand ax maker simultaneously kept two goals active in his conscious awareness (multitasking). As the flakes were removed, he monitored the shape of the core to assess the ongoing progress toward the desired hand ax. In addition, he determined the potential usefulness of the flakes as they were removed and set aside those that might make useful implements. If he noticed that he had just knocked off a really good-looking flake, he momentarily switched his attention from the goal of making a hand ax to one of flake assessment and preservation. Then, once he had evaluated and (in some instances) preserved the flakes, he switched his attention back to making the hand ax. The ability to monitor two processes at once, and to switch back and forth between them, is a form of attentional control not seen in earlier hominin species. It suggests that *Homo heidelbergensis* not only had more control over attention but also had more attentional resources in total compared with their predecessors.

Becoming Human: Fire and Attention

Possibly nothing absorbs and captivates our attention more effectively than fire. Simply staring at the flames can be mesmerizing. Among traditional societies, rituals around fire are commonplace. For example, the !Kung San of southern Africa hold healing dances about every two weeks. They chant and dance around a blazing fire while shaman healers work themselves into a trance (called *kia*) during which they lay their hands on others, transmitting healing power to them. These festive dances are considered essential to the health and vitality of the !Kung community (Katz, 1982). The commonplace nature of communal rituals involving fire suggests that fire played an important role in human cognitive evolution as a venue for social activities. "It is really the beginning of humans. When you have fire, you have people sitting around the campfire together," archaeologist Allison Brooks claims (see Klein & Edgar, 2002, 156).

The first evidence of the possible use of fire is dated to around 1.2 MBYP and was found in the Swartkrans Cave site in South Africa (Brain & Sillen, 1988). This evidence, however, is indirect and controversial, as it consists mainly of the remains of burned, butchered bones. Harvard primatologist Richard Wrangham (2009) contends that at this time, *Homo erectus* was probably cooking meat, which, as a source of protein, would have been essential to the expansion of its brain. Raw meat is hard to digest, a problem that cooking would have solved. There is little doubt that *Homo erectus* was eating more meat, and this would help explain the growth of its neural tissue. However, there are no hearths at Swartkrans, no charcoal or other heat-altered sediments to support the controlled use of fire as a means of cooking. Therefore, the bones might have been burned by a natural source of fire, or they may have been altered chemically.

Two other sites suggest that *Homo erectus* may have been using fire sometime between 1 and 0.8 MBYP. Burned bones and plant remains have been found in sediments dating about one million years ago at

Wonderwerk Cave in southern Africa (Berna et al., 2012). Likewise, burned seeds and wood have been uncovered at Gesher Benot Ya'aqov Cave in Israel, indicating that fire may have been used there at around 800,000 YBP (Alperson-Afil, Richter, & Goren-Ibar, 2007). The remains at Gesher Benot Ya'aqov were found in conjunction with stone tools associated with *Homo erectus*. In addition, evidence of fire use in the form of high proportions of burned bones, possible fire pits, and burned patches of earth can be found at numerous sites such as Zhoukoudian in northern China, Cave of Hearths in southern Africa, Vertesszollos in Hungary, and Bilzingsleben in Germany all dating between 500,000 and 300,000 YBP (Ronen, 1998). A recent review of the European record puts the earliest credible evidence of the controlled use of fire at around 400,000 YBP, based on remains from Schoningen (Germany) and Beeches Pit in England (Roebroeks & Villa, 2011). If stone-lined hearths are considered requisite indicators of the controlled use of fire, then the earliest evidence from caves in both Africa and Eurasia dates to around 250,000 YBP (Klein & Edgar, 2002, 156–157).

By the time that *Homo sapiens* and Neanderthals were present (roughly 200,000 YBP), fire was definitely being used, although each group may have used it differently. Neanderthals used fire extensively. Indeed, they could not have survived the harsh conditions of Ice-Age Europe without it. But their use of fire appears to be entirely pragmatic and relatively transient compared with that of the Cro-Magnons. Some of the best-studied Neanderthal hearths are located at Abric Romani, a rock shelter near Barcelona, Spain, dating to between 70,000 and 40,000 YBP (Vallverdu et al., 2010). Here, numerous hearths were discovered, all indicating small (about one foot across), shallow, short-use, not-very-hot fires. Most are associated with domestic activity such as knapping, butchering, or sleeping. Kebara Cave in Israel has evidence of a few, somewhat larger, and more substantial fires (Bar-Yosef et al., 1992). On the whole, Neanderthal fires tended to be more like those at Abric Romani (for a discussion, see Wynn & Coolidge,

2012, 114–118), as were most of the Cro-Magnons' and even contemporary hunter-gatherers' fires, but not all.

For example, at the Abri Pataud rock shelter site in southwestern France (about 23,000 YBP), Cro-Magnons built large hearths more than three feet in diameter and lined them with stones gathered from a nearby river. Even more impressive are the hearths found at the Dolni Vestonice (23,000 YBP) site in the Czech Republic (Vandiver, Soffer, Klima, & Svoboda, 1989). Hearths of more than six and a half feet in size with deposits more than one and one-third feet thick have been found here. Furthermore, some of the hearths are associated with the shattered remains of carved clay figurines and pellets apparently made to explode when heated. Two Stone-Age "kilns"—clay structures used for firing clay—also have been found, both of which could generate very hot fires. It is clear from Dolni Vestonice that fire had taken on more than just practical significance for the *Homo sapiens* living there. The clay pellets and shattered figurines were likely used in some ceremony involving fire, possibly—as archaeologist Bryan Haden contends—in religious rituals invoking animal spirits (Hayden, 2003, 134–135), that is, the social use of fire as a venue for communal rituals.

Fire, Rituals, and Attention

The social use of fire is relevant in two ways to the evolution of uniquely human levels of attentional control. First, the use of fire as a venue for social activities occurs in conjunction with other archaeological evidence of increased social complexity among our Upper Paleolithic ancestors. The social use of fire was likely both a contributing factor to and an outgrowth of *Homo sapiens'* expanding social world. This is important because interacting more frequently with more people places greater demands on working memory and attentional control.

Cro-Magnon campsites are generally larger, more commonly found, more intensely used and occupied, and (typically) more spatially structured than Neanderthal campsites (Bar-Yosef, 2002; Dickson, 1990, 84–92, 180–189; Hoffecker, 2002, 129, 136; Stringer & Gamble, 1993, 154–158). Many of these sites show evidence of seasonal aggregation, larger population density, and other signs of social complexity and stratification (Mellars, 1996; Vanhaeren & d'Errico, 2005). A larger, more complex social world means interacting more often with more people, many of whom may be only casual acquaintances, which probably put great stress on attention and working memory.

Greater levels of social engagement boost performance on mental tasks involving attention and working memory. An example is a simple working memory task such as counting backward by threes from variously presented numbers. One study found that increased interpersonal activity, such as talking more frequently on the phone or more face-to-face interactions with others, predicted better performance on the counting backward task (Ybarra et al., 2008). Furthermore, ten minutes of social interaction (discussion of a controversial social issue) was as effective as engaging in nonrelational intellectual activity (reading for comprehension, mental rotation tasks, crossword puzzle solving) in boosting performance on mental tasks requiring speed of processing and working memory.

Why is social interaction so cognitively demanding? One study addressed this issue, finding that some social interactions are more important than others in engaging attention and working memory. This study used an attentional allocation task called the "trails task" (Ybarra, Winkielman, Yeh, Burnstein, & Kavanagh, 2011), in which the subject is presented with a sheet containing randomly placed circles with either a number (from 1 to 12) or a letter (from A to M) in each circle. The subject must connect the circles as quickly as possible, switching sequentially back and forth from alphanumeric to numeric order. So, a proper sequence would be A-1, 1-B, B-2, 2-C, and so on.

Before performing the task, the subjects interacted socially with one other person. Some of these social interactions were designed to force the participants to "get into the head" of the other. For example, the subject might try to determine whether the other was lying about something, or she might be told that the other was going to be her partner in an upcoming game to be played. Other social interactions did not have this quality. Only the "getting into the head" type of interactions were found to significantly improve performance on the trails task, for both the cooperative and the competitive interactions. Interactions like these, Ybarra and colleagues argued, demand skill at attentional control because the subject must assess both the nature of the ongoing interaction ("This is going well," "This is really uncomfortable") while simultaneously trying to read the other person's state of mind ("Am I making a good impression?" "Is she trying to be funny?").

These studies show that seemingly simple social interactions are not simple from a cognitive standpoint. Instead, they are quite demanding in regard to the control and allocation of attentional resources. The more frequently we socialize with others, the more we develop and strengthen our attentional skills. All these findings are consistent with neuro-imaging studies showing that social reasoning involves broad areas of the frontal lobe that are used in working memory and executive control. Interestingly, neuro-imaging studies of brain activity while constructing stone tools failed to find activity in these same areas of the frontal lobe. This supports the notion that social factors, more than technological factors, were responsible for the emergence of uniquely human levels of working memory and attention control.

The second way that the social use of fire is relevant to the evolution of attentional control is that fire became a venue for psychologically demanding communal rituals. For example, rituals of initiation are fairly common among traditional societies (Alcorta, 2006; Lutkehaus & Roscoe, 1995). In some cases, these rituals can be quite physically and mentally demanding, requiring, for instance, forced dancing and chanting (to the point of exhaustion), genital mutilation, scarring, and tooth extraction (Glucklich, 2001; Knight, Power, & Watts 1995;

Power, 1998; Whitehouse, 1996). Moreover, the shadowy, grotesque images induced by firelight can heighten the drama, stress, and terror of initiation ceremonies. For example, in the Navajos' initiation by firelight, masked figures emit falsetto screams as they impart sacred knowledge to the initiates (Kluckholn & Leighton, 1974).To endure rituals of this nature requires that the initiates maintain their attentional focus on the ritual activity while suppressing the natural flight or fight response. The initiates' situation is not unlike that of athletes facing a high-pressure moment of success or failure. They must stay focused on the process and not allow the stress of the moment to become paralyzing. Moreover, in our ancestral past, passing tribal initiations was not a trivial matter, as one's status within the tribe, marriageability, and overall prospects for reproductive success were often riding on the outcome.

Rituals involving shamanism also are commonly performed around campfires. Strictly speaking, shamanism is associated with traditional societies across Siberia, Central Asia, and the Saami regions of Scandinavia. However, shamanistic practices of some form have been observed worldwide, in nearly every traditional society (Guenther, 1999; Townsend, 1999; Vitebsky, 2000, 55–56). The term "shaman" comes from the Tungus root *saman*, meaning "one who is excited or raised" or, simply, "to know." This reflects the fact that the shaman's function is to enter an altered state of consciousness in which he or she connects with spiritual forces in order to gain knowledge or cure illness.

The shaman is a spiritual practitioner whose job is to interact with the spiritual world and whose main tool is a ritually induced trance. Achieving the altered state of consciousness associated with shamanism entails focused attention on repetitive motor movements and sensory signals such as dancing, chanting, and drumming, usually done around hypnotic firelight and often aided by psychotropic substances. Shamanism not only is ubiquitous among traditional societies; it also appears to have a deep evolutionary history.

Prehistoric cave art dating as far back as 35,000 YBP provides possible evidence of our Paleolithic ancestors' shamanistic practices (Balter, 2000; Lewis-Williams, 2002; Winkelman, 2002). There also is some evidence that shamanistic practices may have emerged first in Africa, tens of thousands of years before the Upper Paleolithic. In a deep cave site in the Tsodilo Hills of Botswana, Sheila Coulson, an archaeologist at the University of Oslo, reported finding a ritually modified snake-rock, dating around 70,000 YBP (Minkel, 2006). The boulder, measuring 6.5 yards long by 2.2 yards high, has a natural, snakelike appearance that had been intentionally modified so that incoming natural light gave the impression of scales on its surface and firelight gave the impression of undulating movement. For Coulson, these modifications strongly suggest use of the site for consciousness-altering rituals.

Like rituals of initiation, shamanistic rituals were serious business. Along with important group-bonding functions, these rituals also served as a community's primary form of health care (McClenon, 2002; Rossano, 2010a). Those most able to focus their attention to produce a health-enhancing, healing-inducing altered state of mind were also most likely to survive injuries, disorders involving chronic pain, bleeding, and the rigors of childbirth. Accordingly, in the past, the mind was our best defense against sickness, injury, and pain; and social rituals around fire were our primary means of mobilizing that defense.

Getting Into the Flow

As he climbed, Parrado experienced a "loss of self" reminiscent of what he described earlier in the rugby scrum. His concentration became so intensely focused that his very being seemed to disappear as he became only a "pure will to climb" (*Miracle in the Andes*, 197). Never before, he explained, had he felt so focused, driven, and "fiercely alive." In those brilliant moments, all suffering ceased and life became "pure flow."

Sometimes the altered state of consciousness associated with shamanistic rituals reaches such an intense peak that the individual is "lost" in the ecstasy of the moment. During this, he or she may engage in spectacular feats of endurance and pain tolerance. Richard Katz, who extensively studied the !Kung and their healing traditions, describes such a moment involving the shaman-healer Kana who entered the trance state called *kia*:

> Kana continues to walk around in the state of *kia*, like a tightrope walker. He is healing people who are sitting at the little fires on the outskirts of the dance fire. . . . He then goes back to the central dance fire, picks up several reddish-orange coals, and rubs them together in his hands, then over his chest and under his armpits. The sparks fly. He drops the coals back into the fire just as the singers begin to scatter. (Katz, 1982, 71)

The shamans who achieve *kia* told Katz that it is a blissful experience, one of utter transcendence. When in *kia*, they claim, they are more fully alive, more fully themselves than when they are in their normal state of consciousness (Katz, 1982, 43–44). The experience of *kia* appears to be similar to what psychologists refer to as "flow."

In the 1960s, psychologist Mihaly Csikszentmihalyi studied the creative process. While watching painters at work, he noted something profound: when their work was going well, the artists were oblivious to everything outside the creative process. Hunger, fatigue, discomfort— nothing registered in the artists' consciousness in these moments when creation was at its peak. Time seemed to stand still. Hours might pass, but to the artists, they were mere blips of time (Nakamura and Csikszentmihalyi, 2002). He found something similar occurring in others such as chess players, rock climbers, dancers, and surgeons—

any activity in which the participants claimed that the rewards were driven primarily by the joy inherent in the activity itself. Over succeeding decades, Csikszentmihalyi researched this phenomenon, which he termed "flow." Flow, he observed, had the following characteristics:

1. Intense, focused concentration on the activity and the present moment.
2. A total merging of action and awareness, with a loss of reflective self-awareness. In other words, one becomes totally absorbed in the activity. Awareness is fully subsumed in the activity to the point where one loses one's sense of self as a distinct agent. One is the activity.
3. A sense of effortless control in a context of just-manageable challenge. In other words, one's capabilities are maximally stressed, but throughout the process one never feels overwhelmed, incapable, or out of control. Instead, one smoothly and dispassionately adjusts and reacts to the changing circumstances, effectively meeting challenges and obstacles as they arise.
4. A distorted sense of time; where time seems compressed; an hour seems like only moments. Often at the completion of the flow activity, participants are nonplussed at the amount of time that has actually passed.
5. A feeling of intrinsic reward arising from the activity itself. The activity is its own reward with the end result often being secondary. Often during flow, participants claim to be having an optimal experience, their most profound sense of exhilaration, of being "alive." (Nakamura & Csikszentmihalyi 2002, 90)

The flow experience appears to be a human universal, cutting across all distinctions of culture, social class, age, gender, and religion.

Although flow is rarely a matter of life and death, for the expeditionaries, it was, which likely added to the intensity of their experience. As he climbed, Parrado tells us that he lost himself in the activity, that his entire being seemed to merge with the physical act of climbing. He described his body as nothing but a survival vehicle: it was not him; it was his instrument. Finally, he found that in the midst of his prolonged ordeal, he would have glorious, transient moments when his suffering ceased and he was more radiantly alive than he had ever been before.

We have several detailed accounts of death on the mountain. Strong and brave young men who struggled gallantly, who understood life's preciousness and clung to it mightily. But when the fight was over, there was peace, serene resignation. Arturo Nogueira wept as death closed in, not from fear, but because—as he told his friends—he was so close to God. Parrado's climb was a different confrontation with mortality. It is a portrait of life furiously ablaze as it stares at its limit. Some die nobly at peace. Some die more alive than others will ever know.

CHAPTER SEVEN

God of the Mountain

SCENE 10: THE TWO FACES OF GOD

I think there are two types of Gods, one which is shown to you at
the School, sitting in heaven and sending rays to the people who are
on earth, and another one who is the one we knew in the Andes, we
practically lived with him and we asked him [to] help constantly.
(Interview with Roberto Canessa, 2002, http://www.viven.com.uy/
571/eng/EntCanessa012002.asp)

The God you met there [on the mountain] was the same you believed in or saw at the mass?
 No. It was a different God. At school we had the image of an old
bearded God who was walking through the clouds and all the other
things. . . .
 There we met a God who was closer to the indifference of material things, to the humility, to the fact of not having anything. The
more stripped we are of material things, the better we get the figure
of God. (Interview with Carlitos Paez, 2002, http://www.viven.com
.uy/571/eng/EntPaez102002.asp)

Because each survivor of the crash of UAF flight 571 had a somewhat
different experience, it's difficult to make any sweeping generalizations

about lessons learned or lives transformed. More than one survivor, however, reported that the divine presence he encountered on the mountain was quite different from the one discussed in Sunday school. Maybe there are two kinds of gods because are two kinds of religion: natural religion and cultural religion.

In Sunday school, the Andes survivors were introduced to the god of their cultural heritage: the God of the Catholic Church, the God of the Word. On the mountain, they had a direct encounter with the God of the Mountain, the God of the works (of nature). It is the latter God with whom humanity has had the longer relationship. Nature's God is the god of experience, whereas the former is a more recent revelation (or, for some people, an invention). He is the God of theology, of sacred texts. The Andes survivors' encounter with God—an encounter stripped clean of any modern theological reflection—was another instance of their return to ancient ways.

Research on the cognitive basis of religion reveals a suite of mental traits that make religion (not so much theology) natural to us. We are, as psychologist Justin Barrett argues, born believers (Barrett, 2012), although this innate belief is more generally spiritual than specifically religious.

Natural Religion

Over the last two decades or so, scientists have made considerable progress in uncovering the cognitive basis of religion. Far from being something that culture imposes on us, there is considerable evidence that religion is natural to us. That is, our minds are predisposed to think in ways that make certain religious beliefs easy to accept. Most of this research has been done on infants and young children, the goal being to uncover the conceptual "assumptions" that are either innate or easy to establish with minimal experience. The starting point for natural religion is that infants are innately attuned to perceptual features that distinguish animate agents from inanimate objects, and they have different expectations for each.

Agents Bring Order

A twelve-month-old infant is watching a computer display showing a ball going up and over a barrier and coming to rest on the other side next to another ball (Gerley & Csibra, 2003). She watches this event over and over until she becomes bored with it and starts to look away (having become "habituated"). That is, she looks away because she wants to see something new, but what exactly constitutes something new? Suppose now the researcher presents a display of the ball making the same motion (up and over, coming to rest next to the ball on the other side) but without a barrier. Because it's the same motion, shouldn't the infant find this boring as well? The answer is no, the infant "dishabituates"; she starts staring intently at the display again. Maybe it is just because the barrier is missing, which has changed things enough to constitute a novel event. If that were so, then the infant should also dishabituate when the ball takes a direct path across the screen and comes to rest beside the other ball. In this instance, not only is the barrier missing but the path of movement also is different, so the event really ought to be novel. But the infant finds this just as boring as the original display.

The self-generated nature of the ball's movement has caused the infant to assume that it is an animate agent behaving rationally. The ball has a goal—to get near the other ball—and it moves rationally to achieve that goal, by going up and over the barrier. When the barrier was missing and the ball took a direct path, it was simply doing the same thing, moving rationally to achieve its goal. So there was nothing novel in that event. But when the barrier was missing and the ball took the same path as before (up and over), then something strange and novel took place: an animate rational agent achieved its goal in an irrational way, and that is novel.

This research gives us the first religion-relevant assumption that infants use in understanding their world: Self-generated movement indicates the presence of an agent, and an agent is a rational, goal-directed actor. It makes sense that infants would be especially attuned to agents, since it is other people (human agents) who are critical to their survival.

Another study using twelve-month-old infants offers us a second religion-relevant assumption: Human agents are responsible for order, and inanimate objects (or forces) are responsible for disorder (Newman, Keil, Kuhlmeier, & Wynn, 2010). This study used the same looking-time test procedure in which the researchers exposed infants to different events in order to determine which ones were surprising to the infants (and therefore differed from their expectations).

In one event, the infants saw a video of a human hand moving toward an array of blocks that were randomly strewn about. Then an opaque screen dropped so that the infants could not see the hand acting on the blocks. After a pause, the screen lifted to reveal the blocks, which now were neatly arranged. In this instance, the infants viewed an ordering event—the human hand had apparently taken blocks that were randomly strewn about and arranged them neatly. Did the infants find this interesting? No, their looking times showed that this was pretty much what they expected the hand to do.

What they did find surprising was when the event went the other way; that is, the hand seemed to make a neat arrangement into a random mess. Furthermore, the infants had just the opposite expectation for an inanimate object. In another condition, the researchers replaced the human hand with an inanimate, clawlike object. The infants were surprised when the claw turned a random arrangement into a neatly stacked array. Thus, very early in life, infants expect human agents to create order, but inanimate objects and forces (such as the wind) to create disorder.

Another condition produced slightly different results with interesting implications. How would the infants react if the human agent were replaced by a nonhuman one? To test this, Newman and colleagues (2010) showed the infants a circle-face image (a circle with eyes and nose) that created both order and disorder. In neither case did the infants seem surprised. This result suggests that infants expect that nonhuman agents can create either order or disorder.

Agents Are Immortal (in Some Ways)

Humans are imaginative creatures, and children are extraordinarily so. But our imagination is woefully inadequate when imagining our death. What is it like to be dead? The usual answer is that death is like sleep, restful and peaceful. But the reason sleep seems peaceful is that when we awaken, we can compare it to our often hectic daily lives. There is no awakening from death. Thus we cannot compare it with being alive. Death also is often thought of as the end of all suffering and all worry. When someone dies after a prolonged illness, we often assume that he or she is no longer suffering. But this is only true if in death one is able to experience a lack of suffering. If death is truly the end of all experience, then it cannot be the experience of nonsuffering. Instead, it is simply no experiencing at all. But what is it like to "not experience" anything? Here is where our imagination seems to fail us.

Psychologist Jesse Bering has done a number of studies demonstrating the difficulty that people have conceptualizing death. In one study, subjects who self-identified as "extinctivists" (those claiming that all aspects of conscious functioning, including any functions of a "soul," cease upon death) were queried about a fictitious case of a man killed in a car accident (Bering, 2002). About one-third of these subjects' answers indicated that the emotions, desires, and thought processes of the deceased continued beyond death. In other words, even though these subjects explicitly claimed that death was the end of all consciousness, many provided answers that implied the opposite.

If adults have difficultly grappling with the completeness of death, how much more difficult might it be for children? To explore this issue further, researchers presented four- to twelve-year-olds with a rather heartbreaking puppet show in which a mouse, lost in the woods, is eaten by an alligator (Bering & Bjorklund, 2004). The children were then asked a series of questions about the deceased mouse: Does the mouse know that he is no longer alive? Does he want to go home? Can he see where he is? Does he still love his mother? Does his brain still work? Is he still hungry?

Most of the youngest children understood that physiological processes ceased with death and accordingly claimed that the mouse's brain and eyes stopped working after it had been eaten by the alligator. They were less willing, however, to accept that emotions and desires ended with death. Even nearly half the school-age children continued to insist that the mouse still loved his mother and wanted to go home. As children grow older, they increasingly accept that cognitive (does he know that he is no longer alive?) and psychobiological (is he still hungry?) functions cease with death. Bering and Bjorklund (2004) argued that this pattern is actually contrary to what we would expect if children's beliefs in an afterlife were primarily a function of cultural/religious indoctrination. As children mature and have more exposure to religious ideas and education, their afterlife beliefs actually narrow and become less overtly superstitious. For most of them, though, belief in an afterlife never entirely disappears. Indeed, using measures of implicit knowledge, Bering showed that the tendency to assume that some psychological functions persist past death pervades the thinking of most adults (Bering, 2005, 421).

Another, similar study found somewhat different results. Harris and Gimenez (2005) told Spanish seven- to eleven-year-olds different stories about the death of a grandparent. One story used language designed to evoke religious reasoning about death, and another was designed to evoke more biological and secular reasoning. The children were then asked about both the biological and the psychological aspects of mortality. In contrast to Bering and Bjorklund's study (2004), this study found that older children were less willing to accept that all processes ceased upon death, especially for psychological processes and when prompted by the religious narrative. Harris and Gimenez contend that a key difference between their study and Bering and Bjorklund's was that in their study, the deceased was a human, thereby making religious views of the immortality of the soul more directly applicable.

Consistent across all these studies is the ease with which children (and most adults) assume that some aspect of the individual persists beyond death (Barrett, 2012). This assumption probably led to the widespread idea of the soul. Concepts of a soul can be found in some form in nearly all the world's major religions, and they have deep historical roots (Rossano 2010b, 120–121). Even for nonreligious people, the idea of a soul as something distinct from the mind that houses the person's unchanging and potentially immortal "essence" is fairly prevalent (Richart & Harris, 2006; 2008).

Culture very likely uses these natural conceptual predispositions and channels them in particular ways. For example, Bering and his colleagues found that even though Catholic school students showed the same developmental pattern in their afterlife beliefs as did secular students, their tendency to attribute mental and emotional states to dead agents was generally stronger than that of the more secular subjects (Bering, Hernandez-Blasi, & Bjorklund, 2005). Other research found that the Vezo children of rural Madagascar viewed their ancestors' mental functions as being specifically relevant to their cultural role. Thus, the ancestors were more likely to be ascribed specific knowledge states, such as knowing their spouse's name and the location of their house, and specific emotional states, such as missing their children, than general knowledge and emotional states (Astuti & Harris, 2008).

In sum, both children and adults appear to be naturally inclined to think of individuals as defined by an unchanging essence or soul that accounts for their unique personality. Both children and adults find it very hard to envision the complete cessation of all experience that occurs at death. This, in turn, provides a natural entry point for childhood conceptions of immortal souls to make their way into the adult cognition. Even though an adult may explicitly claim to believe that all functions end at death, the idea of a personal essence that persists beyond death makes intuitive sense. In our evolutionary past, when that intuition could serve an adaptive function—as in postulating that an ancestor knows whether tribal traditions are being faithfully followed—this idea had a good chance of being accepted into the adult social context.

The World Is Purposefully Designed

The eminent developmental psychologist Jean Piaget described the world of preschool children as animistic and anthropomorphic (Piaget, 1929). That is, children of this age often endow inanimate and natural objects with the qualities of living organisms (animism) and, more specifically, with human wants and desires (anthropomorphism). For example, when asked why the sun shines or the wind blows, children often respond, "To keep me warm" or "to help me fly my kite." Later research found evidence for this type of thinking across a wide range of different cultures (Looft & Bartz, 1969)

An animistic and anthropomorphic world is one rich with purpose. For young children, everything has a purpose, and often that purpose is directly related to them. Thus, the wind is for "flying my kite," and the sun is to "keep me warm." Psychologist Deborah Kelemen (Kelemen, 1999b) found that four- and five-year-olds see purpose in the natural world as readily as they see it in the human world. For example, when asked what why there are lions in the world, they reply, "For going into zoos." When asked why there are clouds in the sky, they reply, "For raining." This is the same logic that children apply to human-made artifacts: When asked why there are spoons, they reply, "For eating soup." Even when seven- and eight-year-olds are specifically told that adults prefer physical explanations for the properties of natural objects (rocks are pointy because of erosion), the children insist on purposeful or teleological explanations (rocks are pointy so that animals won't sit on them and squash them). In addition, five- to ten-year-old children from both fundamentalist and nonfundamentalist households preferred God as the reason for the emergence of different animal species (Evans, 2001). Similar results have been found for preschoolers in both American and Britain (Kelemen & DiYanni, 2005).

Piaget was probably wrong to assume that this manner of thinking represented an inability to understand the physical aspects of causation. These later studies and others suggest, though, that Piaget was probably right to assert that children are inherently biased toward thinking of the natural world as purposely designed in the same way that human-made artifacts are. In fact, so pervasive is this tendency that Kelemen claims that children are "intuitive theists" (Kelemen, 2004).

Supernatural Agents Know More Than Mere Mortals

Sigmund Freud argued that ideas about God were extensions of ideas about parents. But more recent studies challenge this argument. To demonstrate that children clearly differentiate God's mind from their own and their parents' mind, one study showed children a series of ambiguous visual displays that at first appear to be nothing more than jumbled, confused patterns (Barrett, Newman, & Richert, 2003). Often in displays like these, a critical bit of information ("the blob in the left corner is an eye") leads to the emergence of a meaningful pattern (e.g., a cow). Both before and after being given such disambiguating information, the children were asked whether either their mother or God would understand the displays. Even before they understood the displays themselves, most three- and four-year-olds claimed that God could understand them but that their mothers could not. Furthermore, after the displays were explained to them, most children claimed that their mother now understood them as well. In other words, they confused their mother's knowledge with their own. Most of the children, however, did not change their estimation of what God knew.

Thus the children used neither their mother's knowledge nor their own as a basis for deciding what God knew. This evidence does not support the hypothesis that children use either their parents or themselves as models for constructing their understanding of supernatural agents. Instead, they regard God as the one who knows and perceives everything and who retains this ability even when the children learn more and more about the limitations of other (human) agents.

False-belief tasks are a second way of testing children's understanding of divine minds. In a typical false-belief task, the child is shown a container that is labeled as holding one thing (e.g., crackers) but that when opened is seen to actually hold something else (e.g., pebbles). Children of different ages are then questioned about what others think the container holds. For example, they might be asked, "What will your mother think is in the container?" (assuming that their mother is shown just the outside with the label), or "What will God think is in the container?" The typical result is that younger children (age three or so) tend to think that others know everything that they know. So their mother and other humans will know that there are pebbles in the container, even though the label says "Crackers." Older children (age five or so) recognize that their mother's and other humans' knowledge is limited to the information provided by the label. Consequently, they believe that the container contains crackers and are surprised to find pebbles when it's opened.

Children of all ages, however, claimed that God would know what was in the container (Barrett, Richart, & Driesenga, 2001; Richert & Barrett, 2005). This finding has been replicated cross-culturally, suggesting that it is a general developmental pattern (Knight, Sousa, Barrett, & Atran, 2004). A similar pattern has been observed with perceptual stimuli: as children grow older, they increasingly realize the limitations of human perception while at the same time maintain that God has no limitations. These results suggest that even though an understanding of the limitations of human minds takes time to develop, an understanding of God's mind comes earlier and easier.

From this body of research, we can draw a general outline of our "natural" religious beliefs. We quite readily separate the world into inanimate objects (and forces) and animate agents. Human agents are responsible for order in the world, and inanimate objects and forces typically create disorder. The world is orderly and purposeful, which quite logically suggests the activity of a humanlike supernatural agent. Both natural and supernatural agents have immortal aspects, and supernatural agents have suprahuman knowledge, possibly omniscience.

These appear to be the natural building blocks of religious beliefs. Note, however, that these building blocks do not refer to any particular religion. Instead, what appears to be innate is a general propensity to readily accept and acquire religious ideas, not a specific religion. Our culture then decides how these general cognitive tendencies are molded into a particular religious identity. In this way, religion resembles language. We are born with a general propensity to acquire language, and our culture then dictates the specific language we learn (French, English, Swahili, etc.). Our ancestors should have had these natural building blocks just as we have them today. If so, then we might expect to see these building blocks more clearly in the religious beliefs of traditional societies that have, relative to later global religions, little theological layering.

Ancestor Worship

Two natural building blocks appear in the traditional religious practice of ancestor worship: both the immortality of human agents and the suprahuman knowledge of supernatural agents. If we have trouble accepting that death ends all aspects of a person, then it is not hard to imagine how this led our *Homo sapiens* forebears to conclude that deceased kin and other tribe members were somehow still present. But because they had shed their physical form and were now spiritual, they had access to greater knowledge and power than before.

Ancestor worship is nearly ubiquitous among traditional people (Harvey, 2000; Lee & Daly, 1999), and the elaborate burials of the Upper Paleolithic suggest that ancestor worship may date as far back as thirty thousand years or more. Among traditional people, an elaborate burial with many grave goods usually signals the belief that the deceased is taking his or her place as powerful ancestor (Hayden, 2003, 237) and will continue to be an important player in the tribe's life. He or she will be watching to make sure that traditions are followed and taboos are observed, and if offended, the ancestor will expect a sacrifice as a gesture of atonement.

Animism

Animism, or the belief that the natural world is imbued with spirituality, is another widespread feature of traditional religions (Guenther, 1999, 426; Harvey, 2000; Lee & Daly, 1999, 310). Animism very likely represents an extension of two other natural religious building blocks: our sensitivity to agency and our tendency to see purpose in the world. Whenever infants and children see self-propelled movement, they quickly assign it to an animate creature that has goals and desires, in other words, an agent.

Imagine a child growing up fifty thousand years ago, living daily and intimately with the natural world. Agency was everywhere. Indeed, hunter-gatherers extensively use anthropomorphism when tracking and killing their prey (Gubser, 1965; Marks, 1976; Silberbauer, 1981; see also Barrett, 2005, 456–457; and Mithen, 1996, 168). This strategy involves "getting into the head" of the animal they are trying to capture. What is it thinking? What is it feeling? Where does it want to hide? The animal has goals, desires, and fears just as a human does, and they can be used to predict the animal's movements. Traditional anthropomorphizing, in fact, is as good a hunting strategy as most science-based theories of animal behavior.

The critical point is that in our ancestral past, seeing the animal as a purposeful agent was highly functional. From there, it was only a relatively small conceptual step to seeing the entire natural world as purposeful as well. Those purposes, however, were not always comprehensible to humans nor were they always in the humans' best interests.

Agents of Order and Disorder

The creation myths of traditional societies were reviewed in chapter 3. A curious pattern emerged. Often there was a creator god who sought to establish order in the universe, and often this creator god was humanlike in many ways. But the creator god was frequently foiled by a trickster, who commonly assumed an animal form, with the coyote being especially popular. The humanlike creator and his (or her) animal trickster nemesis can easily be understood as extensions of the infant assumption that human agents create order and nonhuman agents and natural forces create disorder. This assumption intersects neatly with the assumption of purpose in the world. Our ancestors could easily see and appreciate the order and beauty of the natural world. This order had to come from somewhere, and a superhuman agent was a natural explanation. But our ancestors could also see, and frequently were subject to, nature's brutal whimsy. A less-human dis-ordering agent provided a ready explanation for this as well.

The Evolution of the Omniscient God

One feature either missing or deemphasized in traditional religions is the idea of an all-powerful and all-knowing god. For example, the Fang are a traditional people living in Cameroon (West Africa). For them, the earth and the sky are the products of Mebeghe, a great and power-ful creator god. Despite Mebeghe's undisputed grandeur, he is of lit-tle daily concern, and no rituals or sacrifices are made on his behalf (Boyer, 2001). Something similar is true of other traditional hunter-gatherers, such as the !Kung San and Hadza of Africa. Generally, in traditional societies, if there is a single great god at the peak of the divine pyramid, he is far away, detached and unconcerned about mere mortals. Monotheistic ideas—that is, ideas about a single, omniscient and omnipresent God—are relatively recent in the history of religion. Research has given us an explanation of why this might be the case.

Traditional societies are intimate, with a hunter-gatherer band com-posed of a family unit and some extended relatives. These are face-to-face societies in which trust is based on long-standing relationships and kinship ties. But as the human world became more "civilized" and traditional societies were absorbed into chiefdoms and city-states, a serious challenge emerged: How can trust among strangers be estab-lished and maintained? Greater policing is, of course, one answer, and indeed, one study shows that as communities become more diverse, more policing is required to sustain harmony and stability in the community (Kummerli, 2011). This may be a problem, however. Who will police the police? As more layers of oversight are needed, the cost of enforcing cooperation can become prohibitive, and social stability may be lost.

Another solution that some societies use is "offloading" the policing to a supernatural authority (Shariff, Norenzayan, & Henrich, 2009). If we all believe in one, all-knowing, ever-present, and morally concerned god, then we'd better treat everyone fairly or we might get into trouble. This shared belief may be the most cost-effective (and therefore the most adaptive) way of building trust among strangers.

To test for trust among strangers, researchers often use economic games like the investment game. This game is similar to the ultimatum game described in chapter 1, in that it has a proposer who is allowed to allocate some proportion of money to a responder. The experimenter then triples the amount allocated to the responder, and he is then allowed to give some money back to the proposer. Suppose the proposer is given $10 and thoroughly trusts the responder. Then the proposer should give all the money to the responder. When the money is tripled, the responder has $30. The responder then should split the money, giving $15 back to the proposer. The responder ends up with $15 more than what he started with, and the proposer has $5 more. But note that trust is the key here: if the proposer cannot trust the responder to return the money fairly, then she is unlikely to surrender all, or possibly even any, of the initial allotment.

Using the investment game, one study showed that shared religious beliefs increased trust among players (Tan & Vogel, 2008). If the proposer was religious, then she tended to give more money to a religious responder than to a nonreligious one. But if the proposer was not religious, then she gave both religious and nonreligious responders equal amounts of money. In other words, religious people showed greater trust in coreligionists than in nonreligionists. Nonreligious people showed equal trust (or distrust) in religious and nonreligious responders. Thus, shared religious beliefs were used as a signal of who could be trusted. This trust was well placed, as religious responders tended to give back more money to proposers than nonreligious responders did.

Two other studies demonstrate that monotheistic ideas appear to have an advantage over more traditional religious concepts as signals of trust among strangers. The first study looked at why religious people might be both more trusting and trustworthy (Atkinson & Bourrata, 2011), using data from the World Values Survey, which interviewed 350,000 individuals from eighty-seven different countries. These people were asked about their religious beliefs as well as whether certain moral transgressions might be justifiable. The moral transgressions included littering, not paying a bus fare, taking a bribe, and cheating on a spouse. The interviewees were asked whether these activities were always, never, or sometimes justifiable.

The major finding was that religious people across the globe were more unquestioningly dedicated to upholding moral norms. In other words, religious people more often claimed that various moral transgressions were never tolerable. Even more significant was that two particular religious beliefs—belief in a personal god and in heaven and hell—were most strongly related to the intolerance of moral transgression. This is significant because these beliefs are more prominent in monotheistic traditions than in most traditional religions. Thus, if you are a stranger, one way that you might demonstrate to me your trustworthiness is to convincingly demonstrate your monotheistic beliefs. If you do this, I may assume that you are highly intolerant of moral transgression, that you will not lie, cheat, or steal from me because you fear retribution from an omniscient, morally concerned God.

A second study compared the levels of trust of those adhering to world religions with those espousing traditional religions (Henrich et al., 2010). Subjects from fifteen different societies across four continents and Oceania were tested using economic games measuring trust and fairness. The results showed that trust and fairness were significantly related to the degree to which a society adhered to a world religion. That is, those subjects from societies in which world religions were more prominent tended to demonstrate more trust and fairness when playing economic games than did those subjects from societies that practiced traditional religions.

Trust in economic matters is not just playing games. Looking historically at the spread of Islam across Africa, Ensminger (1997) concluded that through its moral beliefs and behavioral standards, Islam served as a mechanism for establishing trust among merchants. Knowing that a stranger was a Muslim was enough on which to build a business relationship. Profits and Islam expanded together.

All this research suggests is that increasing social complexity may have been an important catalyst in the evolution of religious ideas. Larger, more complex societies with greater commerce and trade among strangers required a rethinking of religious beliefs. By lubricating and facilitating commerce, monotheism took hold and spread in more complex, commercially integrated societies. It was a late arrival, however, bringing with it an emphasis on sacred scriptures and a different understanding and experience of the divine (Sanderson & Roberts, 2008).

Mysterium Tremendum

Theologians often distinguish between the book of the works (God's creation, the natural world) and the book of the word (God's revelation to humans, the Bible, Koran). Traditional religions are not book religions (for a discussion of the differences between traditional and book religions, see Hayden 2003, 5–12) but align more closely with both the natural world and our natural religious tendencies. Indeed, before there were religious texts to study, there was nature.

In many traditional societies, a person encountered the divine by going out into nature. Shamans, adolescent initiates, and others in many Native American tribes who sought spiritual wisdom or power often isolated themselves in the wild on "vision quests" in which they communed with guardian spirits (Lewis-Williams, 2002, 166–172). Hilltops, mountain peaks, caves, caverns, and other natural projections and cavities were thought to be access points to the supernatural, the *axis mundi* (Hayden, 2003, 77). Here shamans would enter a ritually induced trance in which they would connect with spiritual forces and acquire special knowledge about healing illness, animal migrations, or the disposition of the gods.

As global religions grew to greater prominence, the raw encounter with the divine so prized by traditional religions was increasingly sidelined. Instead, these global religions emphasized the study of the word, making theology an academic subject, often interconnected with philosophy, history, and even science. Even though sidelined, mystical traditions did not die; the Sufis in Islam, Kabala in Judaism, and Christian mystics such as St. John of the Cross continued to cultivate the direct experience of God. Christian theologian Rudolf Otto described this encounter as *mysterium tremedum et fascinans* (Otto, 1923, 12–41): A great and terrible mystery, an experience of something that is wholly other, frightening and yet at the same time irresistibly attractive. The Andes survivors' encounter with the divine seems to echo Otto's claim.

Nando Parrado described feeling like an intruder on the mountain, as someone who had rudely upset a beautiful and perfect order, undisturbed for millions of years. The survivors' arrival was violent and destructive, and their struggle to survive was crude and chaotic. They had upset a primordial balance, which could be restored only by their removal. Yet despite their interloper status, nearly all felt a benevolent presence. Some of them called it God, and some believed that this God was active in ensuring their survival.

In his speech at the Stella Maris College after their return to Uruguay, survivor Pancho Delgado assured the crowd that God was in the mountains with them. In the morning silence amid the massive peaks, there was, he told them, something "majestic, sensational, and : . . frightening," and they were utterly alone except for the presence of God and his guiding hand (*Alive*, 381). Parrado was less sure of God's presence, but even he acknowledged encountering something great, mysterious, and attractive—something larger than himself—that, in a few exceptional moments, reassured him of the world's ultimate goodness (see *Miracle in the Andes*, 262–263).

The Andes survivors are among those few people who have encountered both the ancient and the recent faces of God. For some, both these faces are simply an illusion, but for others, they must be one and the same, even if they cannot quite understand how this is so. As he lay dying in the Fairchild, Arturo Nogueira chided Parrado for his simplistic views of God. God is incomprehensible, he told him. One should never pray to him for favors, only to get closer to him, to know that he is love (see *Miracle in the Andes*, 84).

In the documentary *Stranded*, Canessa returned to the mountain decades later with his daughter. He described to her how in the midst of their suffering, he felt God's presence. The creator was my friend, he told her. Strangely, however, he did not have that same feeling now. Why not? She asked. Look at all we have, he replied—tents, sandwiches, maps—then, though, we had nothing—*but God*. Maybe ancient mystics risked aloneness before Nature's might because nothing more utterly shattered their hubris. Maybe it was only in weakness and humility that they realized how to live and love more perfectly. For the religious among us, this revelation was seen as a divine gift.

". . . and since we are, therefore I am"

Given their constant struggles with and against Nature, our ancestors were probably more humble than we are today. Humility engendered community, for without one another, they had no chance of surviving. In chapter 1, I discussed the traditional African sense of a communal self, inextricably intertwined with one's family and tribe. This, I argued, was the sense of self known to our ancestors, the sense of self cultivated in rugby and essential to the survival of the stranded passengers of UAF flight 571. Without the "we" of community, the "i" of the individual was doomed.

The story of UAF flight 571 ends with two rituals. The first was Parrado's begging gesture, mentioned in the introduction, with which he established a life-saving link with Catalan. It marked the beginning of the end of the ordeal. But Parrado's gesture was a communal, not an individual, act. Parrado was very nearsighted, and he couldn't see Catalan. Canessa was nearly immobilized from exhaustion, but he had good eyes. He spotted Catalan on horseback staring at them from across the Rio Azufre. Under different circumstances, what followed might easily have been mistaken for slapstick absurdity—Parrado rushing blindly toward the river's edge with Canessa screaming out directions: "go right, right, . . . no, no too far, go left, go left." Two deficient bodies struggling to work as one, each compensating for the other's deficits. The unity of their friendship—symbolized and cemented by the ritual transfer of a personal possession (Panchito's belt)—saved their individual lives. "Since we are, therefore, I am."

The Final Ritual

By the evening of December 22, 1972, all sixteen survivors of the crash of UAF flight 571 had been taken by helicopter to the St. John of God hospital in San Fernando, Chile. The ordeal was over, but the event was not. For *that* to end, one last ritual was required.

For humans, ritual has become a way of taking control of life. Ritual, Catherine Bell tells us, is our way of transforming "physical inevitabili-

The crash site as it looked in 2006, with a cross planted atop a heap of rocks and wreckage.

ties into cultural regularities" (1997, 94). It is how we rob nature of the last word. Biology decides when we are born, but ritual—christening, baptism—decides when we are admitted into our communities. Biology decides when our bodies mature. But ritual—a rite of passage, initiation—decides when we are deemed to be men and women. Biology controls our attractions and lusts, but ritual decides who our legitimate partner will be. Finally, biology terminates our existence, but ritual decides when we are dismissed from the lives of our loved ones.

The laws of physics determined that a plane would crash in the Andes on October 13, 1972, but humans and their rituals brought to an end the event that transpired thereafter. On January 18, 1973, ten members of the Chilean Andean Rescue Corps, a Uruguayan air force officer, and a Catholic priest returned to the crash site. They gathered all the remains and buried them in a sheltered site about a half mile from the Fairchild's fuselage. They erected a rough altar and marked it with an iron cross mounted on it. The priest said a funeral Mass. On the cross was inscribed: "The world to its Uruguayan Brothers. Nearer O God to thee."

Nature runs its course, but we do not stand by as idle observers. We alone take offense at Nature's ignorant indifference, and ritual is the mark we leave testifying to that offense.

Notes

Introduction

1. Technically, it was the copilot, Lieutenant Dante Hector Lugurara, who made the call.

1. Natural Versus Civilized

1. Charles Crawford uses the term "ancestralization" to describe a concept similar to what I am calling "ancient ways." He defines ancestralization as "the process whereby some aspects of a society return to a more ancestral form when ecological, political, religious, or cultural conditions liberalize" (1998, 292). Unfortunately, the term has never gained currency, since it seems to imply a "natural" human nature being artificially masked or distorted by more recent cultural conditions. As I point out here, it is natural for humans to seek to fit in and succeed in their cultures, modern or ancient. But when more recent cultural norms and standards no longer prevail, some ancestral patterns often reassert themselves.

2. This conclusion is reinforced by a study showing that chimpanzees fail to engage in third-party punishment, in which violators of norms are punished by those not directly affected by the violator's actions, something that humans do as well (Riedl, Jensen, Call, & Tomasello, 2012).

4. Mountain Rituals

1. In *Alive* (esp. 174–176), Fito Strauch is portrayed as one of the group's less religious, more skeptical members. He and Pedro Algorta are said to have had

philosophical discussions about God in which both rejected the notion of a "personal" God who intervened in people's destinies. Fito is said to have viewed the nightly rosary as a "sleeping pill" to take their minds off their suffering. After their rescue, he viewed their survival as due mostly to their own efforts, with little direct help from God. There are, however, indications of a spiritual transformation in Fito.

After their rescue, in their hotel in Santiago, Fito and Zerbino had an extended discussion with the Uruguayan Jesuit priest, Father Rodriguez, who had come to speak with the survivors and say Mass the next day. Their discussion continued until five the next morning. Furthermore, In *Miracle in the Andes* (2006, 276), Parrado describes his disagreements with Fito over God's role in their survival. Fito, Parrado tells us, believes that God personally intervened to ensure their survival and may be disappointed that Parrado has not been more public in declaring this message. Their disagreement, however, has not diminished their friendship.

5. Rituals of Love

1. Whether or not it was answered specifically by the Virgin Mary is a judgment I leave to the reader.

2. There is an apparent disagreement in *Alive* and *Miracle in the Andes* on this issue. *Alive* says that Parrado gave Panchito's belt to Canessa with the words "This was a present from Panchito who was my best friend. Now you are my best friend, so you take it" (238). In two places in *Miracle in the Andes*, Parrado clearly states that Canessa took the belt off Panchito's body (267–268) and again in the caption for a photo showing the medics attending to Canessa: "The first medics to arrive at Los Maitenes huddle around Roberto, who is wearing the belt he had taken from the body of Panchito Abal." Although *Alive* is regarded as the canonical account of the Andes survivors' story, I side with Parrado. Either way, the belt works as a ritual object, although in the *Alive* scenario, it simply becomes a ritual gift signifying relational commitment.

6. Ritual Defeats the Mountain

1. See the account in *Harvard University Gazette*, April 18, 2002, http://www
.news.harvard.edu/gazette/2002/04.18/09-tummo.html.

References

Aiello, L. C., & Dunbar, R. I. M. (1993). Neocortex size, group size and the evolution of language. *Current Anthropology, 34*, 184–193.

Alcorta, C. S. (2006). Religion and the life course: Is adolescence an "experience expectant" period for religious transmission? In P. McNamara (Ed.), *Where God and science meet, vol. 2* (55–79). Bridgeport, CT: Praeger.

Alexander, C. N., Rainforth, M. V., & Gelderloos, P. (1991). Transcendental meditation, self-actualization, and psychological health: A conceptual overview and statistical metaanalysis. *Journal of Social Behavior and Personality, 6*, 189–248.

Alexander, R. D. (1989). Evolution of the human psyche. In P. Mellars & C. Stringer (Eds.), *The human revolution: Behavioural and biological perspectives on the origins of modern humans* (455–513). Princeton, NJ: Princeton University Press.

Alperson-Afil, N., Richter, D., & Goren-Inbar, N. (2007). Phantom hearths and the use of fire at Gesher Benot Ya'aqov, Israel. *Paleoanthropology*, 1–15.

Ambrose, S. H. (2002). Small things remembered: Origins of early microlithic industries in sub-Saharan Africa. In R. Elston & S. Kuhn (Eds.), *Thinking small: Global perspectives on microlithic technologies* (9–29). Archaeological Papers of the American Anthropological Society, no. 12.

Ambrose, S. H., & Lorenz, K. G. (1990). Social and ecological models for the Middle Stone Age in Southern Africa. In P. Mellars (Ed.), *The emergence of modern humans* (3–33). Edinburgh: University of Edinburgh Press.

Anastasi, M. W., & Newberg, A. B. (2008). A preliminary study of the acute effects of religious ritual on anxiety. *Journal of Alternative and Complementary Medicine, 14*, 163–165.

Armstrong, D. F., & Wilcox, S. E. (2007). *The gestural origin of language*. Oxford: Oxford University Press.

Astuti, R., & Harris, P. L. (2008). Understanding mortality and the life of the ancestors in rural Madagascar. *Cognitive Science, 32*, 713–740.

References

Atkinson, Q. A., & Bourrata, P. (2011). Beliefs about God, the afterlife and morality support the role of supernatural policing in human cooperation. *Evolution and Human Behavior, 32,* 41–49.

Atran, S. (2002). *In gods we trust.* Oxford: Oxford University Press.

Balter, M. (2000). Paintings in Italian cave may be oldest yet. *Science, 290,* 419–421.

Barclay, L., Aiavao, F., Fenwick, J., & Papua, K. T. (2005). *Midwives tales: Stories of traditional and professional birthing in Samoa.* Nashville, TN: Vanderbilt University Press.

Barendregt, H. (2011). Mindfulness meditation: Deconditioning and changing views. In H. Walach, S. Schmidt, & W. B. Jonas (Eds.), *Neuroscience, consciousness and spirituality* (195–206). New York: Springer.

Barham, L. (2002). Systematic pigment use in the Middle Pleistocene of south-central Africa. *Current Anthropology, 43,* 181–190.

Barnes, M. H. (2000). *Stages of thought.* New York: Oxford University Press.

Baron-Cohen, S., Cox, A., Baird, G., Swettenham, J., Drew, A., Nightingale, N., et al. (1996). Psychological markers of autism at 18 months of age in a large population. *British Journal of Psychiatry, 168,* 158–163.

Barrett, H. C. (2005). Cognitive development and the understanding of animal behavior. In B. J. Ellis & D. F. Bjorklund (Eds.), *Origins of the social mind* (438–467). New York: Guilford Press.

Barrett, J. L. (2012). *Born believers: The science of children's religious beliefs.* New York: Free Press.

Barrett, J. L., Newman, R., & Richert, R. A. (2003). When seeing does not lead to believing: Children's understanding of the importance of background knowledge for interpreting visual displays. *Journal of Cognition and Culture, 3,* 91–108.

Barrett, J. L., & Richert, R. A. (2003). Anthropomorphism or preparedness? Exploring children's God concepts. *Review of Religious Research, 44,* 300–312.

Barrett, J. L., Richert, R. A., & Driesenga, A. (2001). God's beliefs versus mother's: The development of nonhuman agent concepts. *Child Development, 72,* 50–65.

Bar-Yosef, O. (2002). The Upper Paleolithic revolution. *Annual Review of Anthropology, 31,* 363–393.

Bar-Yosef, O., et al. (1992). The excavations in Kebara Cave, Mt. Carmel. *Current Anthropology, 33,* 497–550.

Bell, C. (1997). *Ritual: Perspectives and dimensions.* Oxford: Oxford University Press.

Bennett, L. A., Wolin, S. J., Reiss, D., & Teitelbaum, M. A. (1987). Couples at risk for transmission of alcoholism: Protective influences. *Family Process, 26,* 111–129.

Berger, P. M. (1969). *The social reality of religion.* London: Faber & Faber.

Bering, J. M. (2002). Intuitive conceptions of dead agents' minds: The natural foundations of afterlife beliefs as phenomenological boundary. *Journal of Cognition and Culture, 2,* 263–308.

References

Bering, J. M. (2005). The evolutionary history of an illusion: Religious causal beliefs in children and adults. In B. J. Ellis & D. F. Bjorklund (Eds.), *Origins of the social mind* (411–437). New York: Guilford Press.

Bering, J. M., & Bjorklund, D. F. (2004). The natural emergence of reasoning about the afterlife as a developmental regularity. *Developmental Psychology, 40,* 217–233.

Bering, J. M., Hernández-Blasi, C., & Bjorklund, D. F. (2005). The development of "afterlife" beliefs in secularly and religiously schooled children. *British Journal of Developmental Psychology, 23,* 587–607.

Berna, F. Goldberg, P., Kolska Horwitz, L., Brink, J., Holt, S., Bamford, M., & Chazan, M. (2012). Microstratigraphic evidence of in situ fire in the Acheulean strata of Wonderwerk Cave, South Africa. *Proceedings of the National Academy of Science.* doi: 10.1073/pnas.1117620109.

Bernardi, L., et al. (2001). Effect of rosary prayer and yoga mantras on autonomic cardiovascular rhythms: A comparative study. *British Medical Journal, 323,* 1446–1449.

Bingham, P. M. (1999). Human uniqueness: A general theory. *Quarterly Review of Biology, 74,* 133–169.

Blacking, J. (1973). *How musical is man?* Seattle: University of Washington Press.

Blair, S. L., & Lichter, D. T. (1991). Measuring the division of household labor: Gender segregation of housework among American couples. *Journal of Family Issues, 12,* 91–113.

Blum-Kulka, S. (1997). *Dinner talk: Cultural patterns of socialization in family discourse.* Mahwah, NJ: Erlbaum.

Boddy, J. (2010). The work of Zar: Women and spirit possession in Northern Sudan. In S. Sax, J. Quack, & J. Weinhold (Eds.), *The problem of ritual efficacy* (113–130). Oxford: Oxford University Press.

Boehm, C. (1999). *Hierarchy in the forest.* Cambridge, MA: Harvard University Press.

Boesch, C. (1991). Teaching among wild chimpanzees. *Animal Behaviour, 41,* 530–532.

Boesch, C., & Boesch, H. (1989). Hunting behavior of wild chimpanzees in the Tai National Park. *American Journal of Physical Anthropology, 78,* 547–573.

Bolton, J. M. (1972). Food taboos among the Orang Asli in West Malaysia: A potential nutritional hazard. *American Journal of Clinical Nutrition, 1972, 25,* 789–799.

Bossard, J., & Boll, E. (1950). *Ritual in family living: A contemporary study.* Philadelphia: University of Pennsylvania Press.

Bouldin, P., Bavin, E. L., & Pratt, C. (2002). An investigation of the verbal abilities of children with imaginary companions. *First Language, 22,* 249–264.

Boutcher, S. H., & Crews, D. J. (1987). The effect of preshot attentional routine on a well-learned skill. *International Journal of Sport Psychology, 18,* 30–39.

Boyce, W., Jensen, E., James, S., & Peacock, J. (1983). The family routines inventory: Theoretical origins. *Social Science Medicine, 17,* 193–200.

Boyer, P. (2001). *Religion explained.* New York: Basic Books.

Brain, C. K. (1988). New information from the Swartkrans Cave of relevance to "robust" australopithecines. In F. E. Grine (Ed.), *Evolutionary history of the "robust" australopithecines* (311–324). Hawthorne, NY: Aldine.

Brain, C. K., & Sillen, A. (1988). Evidence from the Swartkrans cave for the earliest use of fire. *Nature, 336,* 464–466.

Brand, R. J. Baldwin, D. A., & Ashburn, L. A. (2002). Evidence for motionese: Modifications in mothers' infant-directed action. *Developmental Science, 5,* 72–83.

Brauer, J., Call, J., & Tomasello, M. (2006). Are apes really inequity averse? *Proceedings of the Royal Society B, 272,* 3123–3128.

Brefczynski-Lewis, J. A., Lutz, A., Schaefer, H. S., Levinson, D. B., & Davidson, R. J. (2007). Neural correlates of attentional expertise in long-term meditation practitioners. *Proceedings of the National Academy of Sciences, 104,* 11483–11488.

Bril, B., Roux, V., & Dietrich, G. (2005). Stone knapping: Khambhat (India), a unique opportunity. In V. Roux & B. Bril (Eds.), *Stone knapping: The necessary conditions for uniquely hominin behaviour* (53–71). Cambridge, UK: McDonald Institute.

Brockelman, W. Y. (1984). Social behavior of gibbons: Introduction. In H. Preuschoft, D. J. Chivers, W. Y. Brockelman, & N. Creel (Eds.), *The lesser apes: Evolutionary and behavioural biology* (285–290). Edinburgh: Edinburgh University Press.

Brody, G. H., & Flor, D. L. (1997). Maternal psychological functioning, family process, and child adjustment in rural, single-parent African American families. *Developmental Psychology, 33,* 1000–1011.

Brosnan, S. F., & de Waal, F. B. M. (2003). Monkeys reject unequal pay. *Nature, 425,* 297–299.

Brosnan, S. F., Shiff, H. C., & de Waal, F. B. M. (2005). Tolerance for inequity may increase with social closeness in chimpanzees. *Proceedings of the Royal Society B, 272,* 253–258.

Brown, S., Merker, B., & Wallin, N. L. (2000). An introduction to evolutionary musicology. In N. L. Wallin, B. Merker, & S. Brown (Eds.), *The origins of music* (3–24). Cambridge, MA: MIT Press.

Bruce, S. (2002). *God is dead: Secularization in the West.* London: Blackwell.

Buck, J. (1988). Synchronized rhythmic flashing in fireflies II. *Quarterly Review of Biology, 112,* 265–289.

Buck, J., & Buck, E. (1978). Toward a functional interpretation of synchronous flashing in fireflies. *American Nature, 112,* 471–492.

Buss, D. (1994). *The evolution of desire*. New York: Basic Books.

Calvin, W. H. (1993). The unitary hypothesis: A common neural circuitry for novel manipulations, language, plan-ahead, and throwing. In K. R. Gibson & T. Ingold (Eds.), *Tools, language, and cognition in human evolution* (230–250). Cambridge: Cambridge University Press.

Carlson, C. A., Bacaseta, P. E., & Simanton, D. A. (1988). A controlled evaluation of devotional meditation and progressive relaxation. *Journal of Psychology and Theology, 16*, 362–368.

Caro, T. M., & Hauser, M. D. (1992). Is there teaching in nonhuman animals? *Quarterly Review of Biology, 67*, 151–174.

Carpenter, J., Burks, S., & Verhoogen, E. (2005). Comparing students to workers: The effects of social framing on behavior in distribution games. In J. Carpenter, G. Harrison, & J. List (Eds.), *Field experiments in economics* (261–290). Greenwich, CT: JAI Press.

Carpenter, M., Akhtar, N., & Tomasello, M. (1998). Fourteen through 18-month-old infants differentially imitate intentional and accidental actions. *Infant Behavior and Development, 21*, 315–330.

Carpenter, M., Tomasello, M., & Striano, T. (2005). Role reversal imitation in 12 and 18 month olds and children with autism. *Infancy, 8*, 253–278.

Carter, O. L., Presti, D. E., Callistemon, C., Ungerer, Y., Lui, G. B., & Pettigrew, J. D. (2005). Meditation alters perceptual rivalry in Buddhist monks. *Current Biology, 15*, R412.

Casler, K., & Kelemen, D. (2005). Young children's rapid learning about artifacts. *Developmental Science, 8*, 472–480.

Caspari, R., & Lee, S.-H. (2004). Older age becomes common late in human evolution. *Proceedings of the National Academy of Sciences, USA, 101*, 10895–10900.

Choi, J.-K., & Bowles, S. (2007). The co-evolution of parochial altruism and war. *Science, 318*, 636–640.

Clark, J. D., & Brown, K. (2001). The Twin Rivers Kopje, Zambia: Stratigraphy, fauna, and artifacts. *Journal of Archaeological Science, 28*, 305–330.

Cohen, D., & Vandello, J. (2001). Honor and "faking" honorability. In R. Nesse (Ed.), *Evolution and the capacity for commitment* (163–185). New York: Russell Sage.

Cohen, E. E. A., Ejsmond-Frey, R., Knight, N., & Dunbar, R. I. M. (2010). Rowers' high-behavioural synchrony is correlated with elevated pain thresholds. *Biology Letters, 6*, 106–108.

Compan, E., Moreno, J., Ruiz, M. T., & Pascual, E. (2002). Doing things together: Adolescent health and family rituals. *Journal of Epidemiology and Community Health, 56*, 89–94.

Connolly, J. A., & Doyle, A.-B. (1984). Relation of social fantasy play to social competence in preschoolers. *Developmental Psychology, 20*, 797–806.

Connor, R. C., Smolker, R., & Bejder, L. (2006). Synchrony, social behaviour and alliance affiliation in Indian Ocean bottlenose dolphins, *Tursiops aduncus*. *Animal Behaviour.* doi: 10.1016/j.anbehav.2006.03.014.

Connors, S. M. (2000). Ecology and religion in Karuk orientations toward the land. In G. Harvey (Ed.), *Indigenous religions: A companion* (139–151). London: Cassell.

Consolmagno, G. (2008). *God's mechanics.* San Francisco: Jossey-Bass.

Coolidge, F. C., & Wynn, T. (2010). *Rise of* Homo sapiens. New York: Wiley.

Corballis, M. C. (2002). *From hand to mouth: The origins of language.* Princeton, NJ: Princeton University Press.

Cosmides, L. (1989). The logic of social exchange: Has natural selection shaped how humans reason? Studies with the Wason selection task. *Cognition, 31,* 187–276.

Crawford, C. (1998). Environments and adaptations: Then and now. In C. Crawford & D. L. Krebs (Eds.), *Handbook of evolutionary psychology* (275–302). Mahwah, NJ: Erlbaum.

Crompton, R. H., Pataky, T. C., Savage, R., D'Août, K., Bennett, M. R., Day, M. H., Bates, K., Morse, S., & Sellers, W. I. (2011). Human-like external function of the foot, and fully upright gait, confirmed in the 3.66 million year old Laetoli hominin footprints by topographic statistics, experimental footprint-formation and computer simulation. *Journal of the Royal Society: Interface,* published online before print, July 20, 2011. doi: 10.1098/rsif.2011.0258.

Cummins, D. D., & Cummins, R. C. (2012). Emotion and deliberative reasoning in moral judgment. *Frontiers in Psychology, 3,* article 328.

Cutler, R. (1975). Evolution of human longevity and the genetic complexity governing aging rate. *Proceedings of the National Academy of Science, USA, 72,* 664–668.

Czech, D. R., Ploszay, A. J., & Burke, K. L. (2004). An examination of the maintenance of preshot routines in basketball free throw shooting. *Journal of Sport Behavior, 27,* 323–329.

Daly, M., & Wilson, M. (1983). *Sex, evolution and behavior.* Boston: Willard Grant.

DeSilva, J. M. (2010). A shift toward birthing relatively large infants early in human evolution. *Proceedings of National Academy of Sciences, 108*(3), 1022–1027. doi/10.1073/pnas.1003865108.

de Waal, F. B. M. (1982). *Chimpanzee politics.* Baltimore: Johns Hopkins University Press.

de Waal, F. B. M. (1988). The communicative repertoire of captive bonobos (*Pan paniscus*) compared to that of chimpanzees. *Behavior, 106,* 183–251.

de Waal, F. B. M. (1990). *Peacemaking among primates.* Cambridge, MA: Harvard University Press.

Dickson, B. D. (1990). *The dawn of belief.* Tucson: University of Arizona Press.

Dissanayake, E. (2000). Antecedents of the temporal arts in early mother-infant interaction. In N. L. Wallin, B. Merker, & S. Brown (Eds.), *The origins of music* (389–410). Cambridge, MA: MIT Press.

Donin, H. H. (1991). *To be a Jew*. New York: Basic Books.

Douglas, M. (1966/1978). *Ritual and purity*. London: Routledge & Kegan Paul.

Dukas, R. (2009). Evolutionary biology of limited attention. In L. Tommasi, M. A., Peterson, & L. Nadel (2009). *Cognitive biology: Evolutionary and developmental perspectives on mind, brain and behavior* (147–161). Cambridge MA.: MIT Press.

Dukas, R., & Ellner, S. (1993). Information processing and prey detection. *Ethology*, *74*, 1337–1346.

Dukas, R., & Kamil, A. C. (2000). The cost of limited attention in blue jays. *Behavioral Ecology*, *11*, 502–506.

Dunbar, R. I. M. (1996). *Grooming, gossip and the evolution of language*. Cambridge, MA: Harvard University Press.

Dunning, E., & Sheard, K. (1979). *Barbarians, gentlemen and players: A sociological study of the development of rugby football*. New York: New York University Press.

Edwards, S. W. (2001). A modern knapper's assessment of the technical skills of the late Acheulean biface workers at Kalambo Falls. In J. D. Clark (Ed.), *Kalambo Falls prehistoric site, vol. 3* (605–611). Cambridge: Cambridge University Press.

Eibl-Eibesfeldt, I. (1975). *Ethology: The biology of behavior*. New York: Holt, Rinehart & Winston.

Ely, R., Gleason, J. B., MacGibbon, A., & Zaretsky, E. (2001). Attention to language: Lessons learned at the dinner table. *Social Development*, *10*(3), 356–373.

Ember, C. R. (1978). Myths about hunter-gatherers. *Ethnology*, *17*, 439–448.

Endicott, P., et al. (2003). The genetic origins of the Andaman Islanders: A vanishing human population. *Current Biology*, *13*, 1590–1593.

Ensminger, J. (1997). Transaction costs and Islam: Explaining conversion in Africa. *Journal of Institutional and Theoretical Economics*, *152*, 4–29.

Erdal, D., & Whiten, A. (1994). On human egalitarianism: An evolutionary product of Machiavellian status escalation? *Current Anthropology*, *35*, 175–183.

Evans, D. W., Gray, F. L., & Leckman, J. F. (1999). The rituals, fears and phobias of young children: Insights from development, psychopathology and neurobiology. *Child Psychiatry and Human Development*, *29*, 261–276.

Evans, E. M. (2001). Cognitive and contextual factors in the emergence of diverse belief systems: Creation versus evolution. *Cognitive Psychology*, *42*, 217–266.

Faisal, A., Stout, D., Apel, J., & Bradley, B. (2010). The manipulative complexity of Lower Paleolithic stone tool making. *PLoS ONE*, *5*, e13718.

Farabaugh, S. M. (1982). The ecological and social significance of duetting. In D. E. Kroodsma, E. H. Miller, & H. Ouellet (Eds.), *Acoustic communication in birds* (85–124). London: Academic Press.

Fehr, E., & Fischbacher, U. (2003). The nature of human altruism. *Nature, 425*, 785–791.

Fein, D. (1976). Just world responding in 6–9 year old children. *Developmental Psychology, 12*, 79–80.

Fiering, C., & Lewis, M. (1987). The ecology of some middle class families at dinner. *International Journal of Behavioral Development, 10*, 377–390.

Fiese, B. H. (1992). Dimensions of family rituals across two generations: Relation to adolescent identity. *Family Process, 31*, 151–162.

Fiese, B. H. (2002). Routines of daily living and rituals in family life: A glimpse of stability and change during the early child-raising years. *Zero to Three, 22*, 10–13.

Fiese, B. H. (2006). *Family routines and rituals.* New Haven, CT: Yale University Press.

Finkelstein, I., & Silberman, N. A. (2001). *The Bible unearthed.* New York: Free Press.

Finkelstein, I., & Silberman, N. A. (2006). *David and Solomon: In search of the Bible's sacred kings and the roots of the Western tradition.* New York: Free Press.

Foley, R., & Lahr, M. (2003). On stony ground: Lithic technology, human evolution, and the emergence of culture. *Evolutionary Anthropology, 12*, 109–122.

Forsythe, R., Horowitz, J. L., Savin, N. E., & Sefton, M. (1994). Fairness in simple bargaining experiments. *Games and Economic Behavior, 6*, 347–369.

Foucart, J., Bril, B., Hirata, S., Morimura, N., Houki, C., Ueno, Y., & Matsuzawa, T. (2005). A preliminary analysis of nut-cracking movements in a captive chimpanzee: Adaptation to the properties of tools and nuts. In V. Roux & B. Bril (Eds.), *Stone knapping: The necessary conditions for uniquely hominin behaviour* (147–157). Cambridge, UK: McDonald Institute.

Franciscus, R. G. (2009). When did the modern human pattern of childbirth arise? New insights from an old Neandertal pelvis. *Proceedings of the National Academy of Sciences, 106*, 9125–9126.

Frecska, E., & Kulcsar, Z. (1989). Social bonding in the modulation of the physiology of ritual trance. *Ethos, 17*, 70–87.

Gayton, W. F., Cielinski, K. L, Francis-Keniston, W. J., & Hearns, J. F. (1989). Effects of pre-shot routine on free throw shooting. *Perceptual and Motor Skills, 68*, 317–318.

Geary, D. C. (2005). *The origin of mind.* Washington, DC: American Psychological Association.

Geissmann, T. (2000). Gibbon songs and human music from an evolutionary perspective. In N. L. Wallin, B. Merker, & S. Brown (Eds.), *The origins of music* (103–123). Cambridge, MA: MIT Press.

Gergely, G., Bekkering, H., & Király, I. (2002). Rational imitation in preverbal infants. *Nature, 415*, 755.

Gergely, G., & Csibra, G. (2003). Teleological reasoning in infancy: The naive theory of rational action. *Trends in Cognitive Science, 7*, 287–292.

References

Ginges, J., Hansen, I., & Norenzayan, A. (2009). Religion and support for suicide attacks. *Psychological Science, 20*, 224–230.

Gleason, J. B., Perlman, R. Y., & Greif, E. B. (1984). What's the magic word: Learning language through politeness routines. *Discourse Processes, 7*, 493–502.

Glucklich, A. (2001). *Sacred pain.* New York: Oxford University Press.

Goodall, J. (1986). *The chimpanzees of Gombe.* Cambridge, MA: Harvard University Press

Goodman, M. J., Estioko-Griffin, A., Griffin, P. B., & Grove, J. S. (1985). Menarche, pregnancy, birth spacing and menopause among the Agta women foragers of Cagayan province, Luzon, the Philippines. *Annals of Human Biology, 12*, 169–177.

Goodnow, J. (1997). Parenting and the transmission and internalization of values: From social-cultural perspectives to within-family analyses. In J. E. Grusec & L. Kuczynski (Eds.), *Parenting and children's internalization of values* (333–361). New York: Wiley.

Goody, J., & Watt, I. (1963). The consequences of literacy. *Comparative Studies in Society and History, 5*, 304–345.

Gould, S. J. (1977). *Ontogeny and phylogeny.* Cambridge, MA: Harvard University Press.

Graefenhain, M., Behne, T., Carpenter, M., & Tomasello, M. (2009). Young children's understanding of joint commitments. *Developmental Psychology, 45*, 1430–1443.

Greenfield, M., & Roizen, I. (1993). Katydid synchronous chorusing in an evolutionary stable outcome of female choice. *Nature, 364*, 618–620.

Grier, R. A., Warm, J. S., Dember, W. N., Matthews, G., Galinsky, T. L., Szalma, J. L., et al. (2003). The vigilance decrement reflects limitations in effortful attention, not mindlessness. *Human Factors, 45*, 349–359.

Gubser, N. J. (1965). *The Nunamiut Eskimos: Hunters of caribou.* New Haven, CT: Yale University Press.

Guenther, M. (1999). From totemism to shamanism: Hunter-gatherer contributions to world mythology and spirituality. In R. B. Lee & R. Daly (Eds.), *Cambridge encyclopedia of hunters and gatherers* (426–433). Cambridge: Cambridge University Press.

Gurven, M., & Kaplan, H. (2007). Longevity among hunter-gatherers: A cross-cultural examination. *Population and Development Review, 33*, 321–365.

Haiman, J. (1994). Ritualization and the development of language. In W. Pagliuca (Ed.), *Perspectives on grammaticalization* (3–28). Amsterdam: John Benjamins.

Harbach, H., Hell, K., Gramsch, C., Katz, N., Hempelmann, G., & Teschemacher, H. (2000). B-endorphin (1–31) in the plasma of male volunteers undergoing physical exercise. *Psychoneuroendocrinology, 25*, 551–562.

Hardman, C. E. (2000). Rites of passage among the Lohorung Rai of East Nepal. In G. Harvey (Ed.), *Indigenous religions* (204–218). London: Cassell.

Harris, P. L. (2000). *The work of the imagination.* London: Blackwell.

Harris, P. L., & Gimenez, M. (2005). Children's acceptance of conflicting testimony: The case of death. *Journal of Cognition and Culture, 5,* 143–164.

Harvey, G. (2000). *Indigenous religions.* London: Cassell.

Hayden, B. (2003). *Shaman, sorcerers, and saints: A prehistory of religion.* Washington, DC: Smithsonian Books.

Heinz, H.-J., and Lee, M. (1979). *Namkwa: Life among the bushmen* Boston: Houghton-Mifflin.

Henrich, J., Boyd, R., Bowles, S., Camerer, C., Fehr, E., Gintis, H., et al. (2001). In search of *Homo economicus*: Behavioral experiments in 15 small-scale societies. *American Economic Review, 91,* 73–78.

Henrich, J., Boyd, R., & Richerson, P. J. (2012). The puzzle of monogamous marriage. *Philosophical Transactions of the Royal Society B, 367,* 657–669.

Henrich, J., Ensminger, J., McElreath, R., Barr, A., Barrett, C., Bolyanatz, A., et al. (2010). Markets, religion, community size, and the evolution of fairness and punishment. *Science, 327,* 1480–1484.

Henrich, J., McElreath, R., Barr, A., Ensminger, J., Barrett, C., Bolyanatz, A., et al. (2006). Costly punishment across human societies. *Science, 312,* 1767–1770.

Hill, K., & Hurtado, A. M. (1996). *Demographic/life history of Ache foragers.* Hawthorne, NY: de Gruyter.

Hill, K., Hurtado, A.M., & Walker, R. S. (2007). High adult mortality among Hiwi hunter-gatherers: Implications for human evolution. *Journal of Human Evolution, 52,* 443–454.

Hobson, P. (2004). *The cradle of thought.* Oxford: Oxford University Press.

Hoffecker, J. F. (2002). *Desolate landscapes: Ice-Age settlement in Europe.* New Brunswick, NJ: Rutgers University Press.

Hoffecker, J. F. (2011). *Landscapes of the mind: Human evolution and the archaeology of thought.* New York: Columbia University Press.

Holloway, R. L., Broadfield, D..C., & Yuan, M. S. (2004). *The human fossil record, vol. 3: Brain endocasts—The paleoneurological evidence.* Hoboken, NJ: Wiley-Liss.

Hove, M. J., & Risen, J. L. (2009). It's all in the timing: Interpersonal synchrony increases affiliation. *Social Cognition, 27,* 949–961.

Howes, C. (1988). Peer interaction of young children. *Monographs of the Society for Research in Child Development, 53*(1), ser. no. 217.

Howlett, T. A., Tomlin, S., Ngahfoong, L., Rees, L. H., Bullen, B. A., Skrinar, G. S., & MacArthur, J. W. (1984). Release of b-endorphin and met-enkephalin during exercise in normal women in response to training. *British Medical Journal, 288,* 295–307.

Hudson, T., & Underhay, E. (1988). *Crystals in the sky: An intellectual odyssey involving the Chumash astronomy, cosmology, and rock art.* Menlo Park, CA: Bellena Press.

Isaac, B. (1987). Throwing and human evolution. *Archaeological Review, 5,* 3–17.

Jacobs, Z., Roberts, R. G., Galbraith, R. F., Deacon, H. J., Grun, R., Macay, A., et al. (2008). Ages for the Middle Stone Age of southern Africa: Implications for human behavior and dispersal. *Science, 322,* 733–735.

Jensen, K., Call, J., & Tomasello, M. (2007). Chimpanzees are rational maximizers in the ultimatum game. *Science, 318,* 107–109.

Jose, P. E. (1990). Just world reasoning in children's imminent justice judgments. *Child Development, 61,* 1024–1033.

Kabat-Zinn, J. (1982). An outpatient program in behavioral medicine for chronic pain patients based on the practice of mindfulness meditation: Theoretical considerations and preliminary results. *General Hospital Psychiatry, 4,* 33–47.

Kahneman, D. (2011). *Thinking fast and slow.* New York: Farrar, Straus & Giroux.

Kapogiannis, D., Barbey, A. K., Su, M., Zamboni, G., Krueger, F., & Grafman, J. (2009). Cognitive and neural foundations of religious belief. *Proceedings of the National Academy of Science.* www.pnas.org_cgi_doi_10.1073_pnas.0811717106.

Katz, R. (1982). *Boiling energy: Community healing among the Kalahari !Kung.* Cambridge, MA: Harvard University Press.

Kaul, P., Passafiume, J., Sargent, R. C., & O'Hara, B. F. (2010). Meditation acutely improves psychomotor vigilance, and may decrease sleep need. *Behavioral and Brain Functions, 6,* 47.

Kelemen, D. (1999a). Beliefs about purpose: On the origins of teleological thought. In M. C. Corballis & S. E. G. Lea (Eds.), *The descent of mind* (278–294). Oxford: Oxford University Press.

Kelemen, D. (1999b). Why are rocks pointy? Children's preference for teleological explanations of the natural world. *Developmental Psychology, 35,* 1440–1453.

Kelemen, D. (2004). Are children "intuitive theists"? Reasoning about purpose and design in nature. *Psychological Science, 15,* 295–301.

Kelemen, D., & DiYanni, C. (2005). Intuitions about origins: Purpose and intelligence in children's reasoning about nature. *Journal of Cognition and Development, 6,* 3–31.

Keller, H., Scholmerich, A., & Eibl-Eibesfeldt, I. (1988). Communication patterns in adult-infant interactions in Western and non-Western cultures. *Journal of Cross-Cultural Psychology, 19,* 427–445.

Kelly, R. C. (1985). *The Nuer conquest: The structure and development of an expansionist system.* Ann Arbor: University of Michigan Press.

Keltner, B. (1990). Family characteristics of preschool social competence among black children in a Head Start program. *Child Psychiatry and Human Development, 21,* 95–108.

Keltner, B., Keltner, N. L., & Farren, E. (1990). Family routines and conduct disorders in adolescent girls. *Western Journal of Nursing Research, 12,* 161–174.

References

Kenward, B., Karlsson, M., & Persson, J. (2010). Over-imitation is better explained by norm learning than by distorted causal learning. *Proceedings of the Royal Society B: Biological Science.* doi: 10.1098/rspb.2010.1399.

Klein, R. G. (1969). *Man and culture in the Pleistocene: A case study.* San Francisco: Chandler.

Klein, R. G. (2005). Hominin dispersal in the Old World. In C. Scarre (Ed.), *The human past: World prehistory and the development of human societies* (84–123). London: Thames & Hudson.

Klein, R. G., & Edgar, B. (2002). *The dawn of human culture.* New York: Wiley.

Klima, B. (1988). A triple burial from the upper Paleolithic of Dolni Vestonice, Czechoslovakia. *Journal of Human Evolution, 16*, 831–835.

Kluckholn, C., & Leighton, D. (1974). *The Navaho.* Cambridge, MA: Harvard University Press.

Knight, C. D., Power, C., & Watts, I. (1995). The human symbolic revolution: A Darwinian account. *Cambridge Archaeological Journal, 5*, 75–114.

Knight, N., Sousa, P., Barrett, J., & Atran, S. (2004). Children's attributions of beliefs to humans and God: Cross-cultural evidence. *Cognitive Science, 28*, 117–126.

Kohler, W. (1927). *The mentality of apes.* New York: Harcourt Brace.

Kohn, M., & Mithen, S. (1999). Handaxes: Products of sexual selection? *Antiquity, 73*, 518–526.

Konner, M. (2010). *The evolution of childhood.* Cambridge, MA: Harvard University Press.

Kummerli, R (2011). A test of evolutionary policing theory with data from human societies. *PLoS ONE, 6* (9): e24350. doi:10.1371/journal.pone.0024350.

Lalonde, C. E., & Chandler, M. J. (1995). False belief understanding goes to school: On the social-emotional consequences of coming early or late to a first theory of mind. *Cognition and Emotion, 9*, 167–185.

Lawick-Goodall, J. (1968). The behavior of free-living chimpanzees in the Gombe Stream Reserve. *Animal Behavior Monographs, 1*, 161–311.

Lee, R. B., & Daly, R. (1999). Introduction: Foragers and others. In R. B. Lee & R. Daly (Eds.), *The Cambridge encyclopedia of hunters and gatherers* (1–22). Cambridge: Cambridge University Press.

Leeming, D. A. (2010). *Creation myths of the world: An encyclopedia.* Santa Barbara, CA: ABC-CLIO.

Leonard, H. L., Goldberger, E. L., Rapoport, J. L., Cheslow, D. L.. & and Swedo, S. E. (1990). Childhood rituals: Normal development or obsessive compulsive symptoms? *Journal of the American Academy of Child and Adolescent Psychiatry, 29*, 17–23.

Lepre, C. J., Roche, H., Kent, D. V., Harmand, S., Quinn, R. L., Brugal, J.-P., et al. (2011). An earlier origin for the Acheulian. *Nature, 477*, 82–85.

References

Lerner, M. J., Somers, D. G., Reid, D., Chiriboga, D., & Tierney, M. (1991). Adult children as caregivers: Egocentric biases in judgments of sibling contributions. *The Gerontologist, 31*, 746–755.

Levenson, R. W. (2003). Blood, sweat, and fears: The autonomic architecture of emotion. In P. Ekman, J. J. Campos, R. J. Davidson, & F. B. M. de Waal (Eds.), *Emotions inside out* (348–366). New York: New York Academy of Sciences.

Lewis-Williams, J. D. (2002). *The mind in the cave.* London: Thames & Hudson.

Lidor, R., & Singer, R. N. (2000). Teaching preperformance routines to beginners. *Journal of Physical Education, Recreation, & Dance, 71*, 34–36.

Lienard, P., & Boyer, P. (2006). Whence collective rituals: A cultural selection model of ritualized behavior. *American Anthropologist, 108*, 814–824.

Litovsky, R. Y., Colburn, H. S., Yost, W. A., & Guzman, S. J. (1999). The precedence effect. *Journal of the Acoustical Society of America, 106*, 1633–1654.

Lobmeyer, D. L., & Wasserman, E. A. (1986). Preliminaries to free throw shooting: Superstitious behaviour? *Journal of Sports Behaviour, 9*, 70–78.

Lonsdale, C., & Tam, J. T. M. (2007). On the temporal and behavioural consistency of pre-performance routines: An intra-individual analysis of elite basketball players' free throw shooting accuracy. *Journal of Sports Sciences, 26*, 259–266.

Looft, W. R., & Bartz, H. (1969). Animism revisited. *Psychological Bulletin, 7*, 1–19.

Lutkehaus, N. C., & Roscoe, P. B. (1995). *Gender rituals: Female initiation in Melanesia.* London: Routledge.

Lutz, A., Slagter, H. A., Dunne, J. D., & Davidson, R. J. (2008). Attention regulation and monitoring in meditation. *Trends in Cognitive Sciences, 14*, 163–169.

Lyons, D. E., Young, A. G., & Keil, F. C. (2007). The hidden structure of overimitation. *Proceedings of the National Academy of Science, USA, 104*, 19751–19756.

MacDonald, K. (1995). The establishment and maintenance of socially imposed monogamy in Western Europe. *Politics and Life Science, 14*, 3–23.

Mackworth, N. H. (1948). The breakdown of vigilance during prolonged visual search. *Quarterly Journal of Experimental Psychology, 1*, 6–21.

Malinowski, B. (1922). *Argonauts of the western Pacific: An account of native enterprise and adventure in the archipelagoes of Melanesian New Guinea.* London: Routledge.

Malinowski, B. (1929). *The sexual life of savages in north-western Melanesia: An ethnographic account of courtship, marriage and family life among the natives of the Trobriand Islands, British New Guinea.* New York: H. Ellis.

Mar, R. A., Oatley, K., Hirsh, J., dela Paz, J., & Peterson, J. B. (2006). Bookworms versus nerds: Exposure to fiction versus non-fiction, divergent associations with social ability, and the simulation of fictional social worlds. *Journal of Research in Personality, 40*, 694–712.

Marchant, L. F., & McGrew, W. C. (2005). Percussive technology: Chimpanzee bao-bab smashing and the evolutionary modeling of hominin knapping. In V. Roux & B. Bril (Eds.), *Stone knapping: The necessary conditions for uniquely hominin behaviour* (341–350). Cambridge, UK: McDonald Institute.

Marks, S. A. (1976). *Large mammals and brave people: Subsistence hunters in Zambia.* Seattle: University of Washington Press.

Markson, S., & Fiese, B. H. (2000). Family rituals as a protective factor against anxiety for children with asthma. *Journal of Pediatric Psychology, 25,* 471– 479.

Martens, S., Munneke, J., Smid, H., & Johnson, A. (2006). Quick minds don't blink: Electrophysiological correlates of individual differences in attentional selection. *Journal of Cognitive Neuroscience, 18,* 1423–1438.

Martini, M. (2002). How mothers in four American cultural groups shape infant learning during mealtime. *Zero to Three, 22,* 14–20.

Marwell, G., & Ames, R. E. (1981). Economists free-ride: Does anyone else? *Journal of Public Economics, 15,* 295–310.

Marzke, M. W. (1996). Evolution of the hand and bipedality. In A. Lock & C. R. Peters (Eds.), *Handbook of symbolic evolution* (126–154). Oxford: Oxford University Press.

Matsuzawa, T. (2001). Primate foundations of human intelligence: A view from tool use in nonhuman primates and fossil hominids. In T. Matsuzawa (Ed.), *Primate origins of human cognition and behavior* (3–25). New York: Springer.

Mbiti, J. (1970). *African religions and philosophies.* New York: Doubleday

McBrearty, S., & Tryon, C. (2006). From Acheulean to Middle Stone Age in the Kapthurin formation. In E. Hovers & S. L. Kuhn (Eds.), *Transitions before the transition* (257–277). New York: Springer.

McClenon, J. (2002). *Wondrous healing: Shamanism, human evolution and the origin of religion.* DeKalb: Northern Illinois University Press.

McKinney, M. L., & McNamara, K. J. (1991). *Heterochrony: The evolution of ontogeny.* New York: Plenum.

McNeill, W. H. (1995). *Keeping together in time: Dance and drill in human history.* Cambridge, MA: Harvard University Press.

Mellars, P. (1996). *The Neanderthal legacy.* Princeton, NJ: Princeton University Press.

Menkiti, I. A. (1984). Person and community in African traditional thought. In R. A. Wright (Ed.), *African philosophy. An introduction* (171–181). Lanham, MD: University Press of America.

Miller, D. T., & Ross, M. (1975). Self-serving biases in the attribution of causality: Fact or fiction? *Psychological Bulletin, 82,* 213–225.

Minkel, J. R. (2006). Offerings to a stone snake provide the earliest evidence of religion. *Scientific American* online: www.sciam.com/article.cfm?articleID=3FE89 A86-E7F2–99DF-366D045A5BF3EAB1 (accessed 5/21/07).

Mitani, J. C. (1985). Gibbon song duets and intergroup spacing. *Behavior, 92*, 59–96.

Mitani, J. C., & Brandt, K. L. (1994). Social factors affect the acoustic variability in the long-distance calls of male chimpanzees. *Ethology, 96*, 233–252.

Mithen, S. J. (1996). *A prehistory of the mind: A search for the origins of art, religion, and science*. London: Thames & Hudson.

Mithen, S. J. (2006). *The singing Neanderthals: The origins of music, language, mind, and body*. Cambridge, MA: Harvard University Press.

Murray, L., & Trevarthen, C. (1986). The infant's role in mother-infant communication. *Journal of Child Language, 13*, 15–29.

Nadel, J., Carchon, I., Kervella, C., Marcelli, D., & Reserbat-Plantey, D. (1999). Expectancies for social contingency in 2-month-olds. *Developmental Science, 2*, 164–173.

Nagy, E., & Molnar, P. (2004). Homo imitans or homo provocans? The phenomenon of neonatal imitation. *Infant Behavior and Development, 27*, 57–63.

Nakamura, J., & Csikszentmihalyi, M. (2002). The concept of flow. In C. R. Snyder & S. J. Lopez (Eds.), *The handbook of positive psychology* (89–92). Oxford: Oxford University Press.

Nauright, J., & Chandler, T. J. L. (1996). *Making men: Rugby and masculine identity*. London: Frank Cass.

Nemeroff, C., & Rozin, P. (1994). The contagion concept in adult thinking in the United States: Transmission of germ and interpersonal influence. *Ethos: Journal for the Society of Psychological Anthropology, 22*, 158–186.

Newman, G. E., Keil, F. C., Kuhlmeier, V. A., & Wynn, K. (2010). Early understandings of the link between agents and order. *Proceedings of the National Academy of Science, 107*, 17140–17145.

Nielsen, M., & Tomaselli, K. (2010). Overimitation in Kalahari Bushman children and the origins of human cultural cognition. *Psychological Science, 21*, 729–736.

Norton, D. G. (1993). Diversity, early socialization, and temporal development: The dual perspective revisited. *Social Work, 38*, 82–90.

Oatley, K. (1999). Why fiction may be twice as true as fact: Fiction as cognitive and emotional simulation. *Review of General Psychology, 3*, 101–117.

Oberg, K. (1973). *The social economy of the Tlinglit Indians*. Seattle: University of Washington Press.

O'Halloran, J. P., Jevning, R., Wilson, A. F., Skowsky, R., Walsh, R. N., & Alexander, C. (1985). Hormonal control in a state of decreased activation: Potentiation of arginine vasopressin secretion. *Physiology and Behavior, 35*, 591–595.

Ong, W. (1982). *Orality and literacy: The technologizing of the world*. New York: Methuen.

Otto, R. (1923). *The idea of the holy*. Oxford: Oxford University Press.

Parfitt, S., & Roberts, M. (1999). Human modification of faunal remains. In M. B. Roberts & S. Parfitt (Eds.), *Boxgrove, A Middle Pleistocene hominid site at Eartham Quarry, Boxgrove, West Sussex* (395–415). London: English Heritage.

Parker, S. T., & McKinney, M. L. (1999). *Origins of social intelligence.* Baltimore: Johns Hopkins University Press.

Parrado, N. (2006). *Miracle in the Andes.* New York: Three Rivers Press.

Paton, W. R. (1912). The armour of Achilles. *The Classical Review, 26,* 1–4.

Pelegrin, J. (2005). Remarks about archaeological techniques and methods of knapping: Elements of a cognitive approach to stone knapping. In V. Roux & B. Bril (Eds.), *Stone knapping: The necessary conditions for uniquely hominin behaviour* (23–33). Cambridge, UK: McDonald Institute.

Pelegrin, J. (2009). Cognition and the emergence of language: A contribution from lithic technology. In S. A. de Beaune, F. L. Coolidge, & T. Wynn (Eds.), *Cognitive archaeology and human evolution* (95–108). Cambridge: Cambridge University Press.

Phelps, K. E., & Woolley, J. D. (1994). The form and function of young children's magical belief. *Developmental Psychology, 30,* 385–394.

Phillips, J. (1996). The hard man: Rugby and the formation of male identity in New Zealand. In J. Nauright & T.J. L. Chandler (Eds.), *Making men: Rugby and masculine identity* (70–90). London: Frank Cass.

Piaget, J. (1929). *The child's conception of the world.* New York: Harcourt Brace.

Pietrowsky, R., Braun, D., Fehm, H. L., Pauschinger, P., & Born, J. (1991). Vasopressin and oxytocin do not influence early sensory processing but affect mood and activation in man. *Peptides, 12,* 1385–1391.

Plooij, F. X. (1978). Tool use during chimpanzee's bush pig hunt. *Carnivore, 1,* 103–106.

Power, C. (1998). Old wives' tales: The gossip hypothesis and the reliability of cheap signals. In J. R. Hurford, M. Studdert-Kennedy, & C. Knight (Eds.), *Approaches to the evolution of language: Social and cognitive bases* (111–129). Cambridge: Cambridge University Press.

Pruetz, J. D., & Bertolani, P. (2007). Savanna chimpanzees (*Pan troglodytes verus*) hunt with tools. *Current Biology, 17,* 412–417.

Radcliff-Brown, A. R. (1952). *Structure and function in primitive society.* New York: Free Press.

Rakoczy, H. (2008). Taking fiction seriously: Young children's understand the normative structure of joint pretend games. *Developmental Psychology, 44,* 1195–1201.

Rakoczy, H., Brosche, N., Warneken, F., & Tomasello, M. (2009). Young children's understanding of the context-relativity of normative rules in conventional games. *British Journal of Developmental Psychology, 27,* 445–456.

Rakoczy, H., Warneken, F., & Tomasello, M. (2008). Sources of normativity: Young children's awareness of the normative structure of games. *Developmental Psychology, 44,* 875–881.

Ramey, S. L., & Juliusson, H. K. (1998). Family dynamics at dinner: A natural context for revealing basic family processes. In M. Lewis & C. Feiring (Eds.), *Families, risk, and competence* (31–52). Mahwah, NJ: Erlbaum.

Rappaport, R. A. (1999). *Ritual and religion and the making of humanity*. Cambridge: Cambridge University Press.

Rawls, J. (1971). *A theory of justice*. Cambridge, MA: Harvard University Press.

Read, P. P. (1974/2002). *Alive: The story of the Andes survivors*. New York: Harper Perennial.

Reaney, L. T., Sims, R. A., Sims, S. W. M., Jennions, M. D., & Backwell, P. R. Y. (2008). Experiments with robots explain synchronized courtship in fiddler crabs. *Current Biology, 18*, R62–63.

Reddy, V. (2008). *How infants know minds*. Cambridge, MA: Harvard University Press.

Reed, D. L., Light, J. E., Allen, J. M., & Kirchman, J. J. (2007). Pair of lice lost or parasites regained: The evolutionary history of anthropoid primate lice. *BMC Biology, 5*, 7.

Richerson, P. J., & Boyd, R. (2001). The evolution of subjective commitment to groups: A tribal instincts hypothesis. In R. M. Nesse (Ed.), *Evolution and the capacity for commitment* (186–220). New York: Russell Sage.

Richerson, P. J., & Boyd, R. (2005). *Not by genes alone: How culture transformed human evolution*. Chicago: University of Chicago Press.

Richert, R. A., & Barrett, J. L. (2005). Do you see what I see? Young children's assumptions about God's perceptual abilities. *International Journal for the Psychology of Religion, 15*, 283–295.

Richert, R. A., & Harris, P. L. (2006). The ghost in my body: Children's developing concept of the soul. *Journal of Cognition and Culture, 6*, 409–427.

Richert, R. A., & Harris, P. L. (2008). Dualism revisited: Body *vs.* mind *vs.* soul. *Journal of Cognition and Culture, 8*, 99–115.

Rieldl, K., Jensen, K., Call, J., & Tomasello, M. (2012). No third-party punishment in chimpanzees. *Proceedings of the National Academy of Sciences, 109*, 14824–14829.

Roby, A. C., & Kidd, E. (2008). The referential communication skills of children with imaginary companions. *Developmental Science, 11*, 531–540.

Roche, H. (2005). From simple flaking to shaping: Stone knapping evolution among hominins. In V. Roux & B. Bril (Eds.), *Stone knapping: The necessary conditions for uniquely hominin behaviour* (35–48). Cambridge, UK: McDonald Institute.

Roebroeks, W., & Villa, P. (2011). On the earliest evidence of habitual fire use in Europe. *Proceedings of the National Academy of Sciences*. www.pnas.org/cgi/doi/10.1073/pnas.1018116108.

Rogers, A. R., Iltis, D., & Wooding, S. (2004). Genetic variation at the MCIR Locus and the time since loss of human body hair. *Current Anthropology, 45*(1), 105–24.

Ronen, A. (1998). Domestic fire as evidence for language. In T. Akazawa, K. Aoki, & O. Bar-Yosef (Eds.), *Neanderthals and modern humans in Western Asia* (439–447). New York: Plenum.

References

Rosenberg, K. R., & Trevathan, W. (2002). Birth, obstetrics and human evolution. *BJOG: An International Journal of Obstetrics and Gynaecology, 109*, 1199–1206.

Ross, H. S., & Lollis, S. P. (1987). Communication within infant social games. *Developmental Psychology, 23*, 241–248.

Ross, M., & Sicoly, F. (1979). Egocentric biases in availability and attribution. *Journal of Personality and Social Psychology, 37*, 322–336.

Rossano, M. J. (2010a). Harnessing the placebo effect: Religion as a cultural adaptation. In U. Frey (Ed.), *The nature of God: Evolution and religion* (111–128). Berlin: Tectum-Verlag.

Rossano, M. J. (2010b). *Supernatural selection: How religion evolved.* Oxford: Oxford University Press.

Rossano, M. J. (2012). The essential role of ritual in the transmission and reinforcement of social norms. *Psychological Bulletin, 138*, 529–549. doi: 10.1037/a0027038.

Roth, A. E., Prasnikar, V., Okuno-Fujiwara, M., & Zamir, S. (1991). Bargaining and market behavior in Jerusalem, Ljubljana, Pittsburgh, and Tokyo: An experimental study. *American Economic Review, 81*, 1068–1095.

Roux, V., & David, E. (2005). Planning abilities as a dynamic perceptual-motor skill: An actualist study of different levels of expertise involved in stone knapping. In V. Roux & B. Bril (Eds.), *Stone knapping: The necessary conditions for uniquely hominin behaviour* (91–108). Cambridge, UK: McDonald Institute.

Ruff, C. B., Trinkaus, E., & Holliday, T. W. (1997). Body mass and encephalization in Pleistocene *Homo. Nature, 387*, 173–176.

Sanderson, S. K., & Roberts, W. W. (2008). The evolutionary forms of the religious life: A cross-cultural, quantitative analysis. *American Anthropologist, 110*, 454–466.

Sartre, J.-P. (1956). *Being and Nothingness: An essay on phenomenological ontology* (trans. with an introduction by H. E . Barnes). New York: Philosophical Library.

Schelde, T., & Hertz, M. (1994). Ethology and psychotherapy. *Ethology and Sociobiology, 15*, 383–392.

Schele, L., & Freidel, D. (1990). *A forest of kings.* New York: Morrow.

Schick, K. (1994). The Movius line reconsidered: Perspectives on the earlier Palaeolithic of Eastern Asia. In R. S. Corruccini & R. L. Ciochon (Eds.), *Integrative paths to the past: Paleoanthropological advances in honor of F. Clark Howell* (569–595). Englewood Cliffs, NJ: Prentice Hall.

Schick, K., & Toth, N. (1993). *Making silent stones speak.* New York: Simon & Schuster.

Schieffelin, E. (1976). *The sorrow of the lonely and the burning of the dancers.* New York: St. Martin's Press.

Schjoedt, U., Stødkilde-Jørgensen, H., Geertz, A. W., & Roepstorf, A. (2009). Highly religious participants recruit areas of social cognition in personal prayer. *Social Cognitive and Affective Neuroscience, 4*, 199–207.

Schmidt, M. F. H., Rakoczy, H., & Tomasello, M. (2011). Young children attribute normativity to novel actions without pedagogy or normative language, *Developmental Science, 14*, 530–539.

Schuck, L. A., & Bucy, J. E. (1997). Family rituals: Implications for early intervention. *Topics in Early Childhood Special Education, 17*, 477–493.

Schwier, C., van Maanen, C., Carpenter, M., & Tomasello, M. (2006). Rational imitation in 12-month-old Infants. *Infancy, 10*, 303–311.

Seaton, E. K., & Taylor, R. D. (2003). Exploring familial processes in urban, low-income African American families. *Journal of Family Issues, 24*, 627–644.

Seiffge-Krenke, I. (1993). Close friendship and imaginary companions in adolescence. *New Directions for Child Development, 60*, 73–87.

Seiffge-Krenke, I. (1997). Imaginary companions in adolescence: Sign of a deficient or positive development. *Journal of Adolescence, 20*, 137–154.

Semaw, S., Renne, P., Harris, J. W. K., Feibel, C. S., Bernor, R. L., Fesseha, N., & Mowbray, K. (1997). 2.5-million-year-old stone tools from Gona, Ethiopia. *Nature, 385*, 333–336.

Shapiro, K. L., Arnell, K. A., & Raymond, J. E. (1997). The attentional blink. *Trends Cognitive Sciences, 1*, 291–296.

Shariff, A. F., & Norenzayan, A. (2007). God is watching you: Priming God concepts increases prosocial behavior in an anonymous economic game. *Psychological Science, 18*, 803–809.

Shariff, A. F., Norenzayan, A., & Henrich, J. (2009). The birth of high gods: How the cultural evolution of supernatural policing agents influenced the emergence of complex, cooperative human societies, paving the way for civilization. In M. Schaller, A. Norenzayan, S. Heine, T. Yamagishi, & T. Kameda (Eds.), *Evolution, culture and the human mind* (117–136). New York: Taylor-Francis Group.

Shostak, M. (1981). *Nisa: The life and words of a !Kung woman.* Cambridge, MA: Harvard University Press.

Silberbauer, G. (1981). *Hunter and habitat in the central Kalahari Desert.* Cambridge: Cambridge University Press.

Slagter, H.A., Lutz, A., Greischar, L.L., Francis, A.D., Nieuwenhuis, S., et al. (2007). Mental training affects distribution of limited brain resources. *PLOS Biology, 5*(6), e138. doi:10. 1371/journal.pbio.0050138.

Smuts, B. B., & Watanabe, J. M. (1990). Social relationships and ritualized greetings in adult male baboons (*Papio cynocephalus anubis*). *International Journal of Primatology, 11*, 147–172.

Soltis, J., Boyd, R., & Richerson, P. J. (1995). Can group-functional behaviors evolve by cultural group selection? An empirical test. *Current Anthropology, 36*, 473–494.

Sosis, R. (2004). The adaptive value of religious ritual. *American Scientist, 92*, 166–172.

Southard, D., & Amos, B. (1996). Rhythmicity and preperformance ritual: Stabilizing a flexible system. *Research Quarterly for Exercise and Sport, 67,* 288–296.

Spagnola, M., & Fiese, B. H. (2007). Family routines and rituals: A context for development in the lives of young children. *Infants & Young Children, 20,* 284–299.

Stanton, M. E. (1979). The myth of natural childbirth. *Journal of Nurse-Midwifery, 24,* 25–29.

Stark, R. (2003). *For the glory of God.* Princeton, NJ: Princeton University Press.

Stark, R. (2006). *The victory of reason: How Christianity led to freedom, capitalism, and Western success.* New York: Random House.

Steiner, F. B. (1956/2004). *Taboo.* London: Routledge.

Stephens, D. W., & Krebs, J. R. (1986). *Foraging theory.* Princeton, NJ: Princeton University Press.

Sterelny, K. (1996). The return of the group. *Philosophy of Science, 63,* 562–584.

Stern, K., & McClintock, M. K. (1998). Regulation of ovulation by human pheromones. *Nature, 392,* 177–179.

Stout, D. (2005). The social and cultural context of stone-knapping skill acquisition. In V. Roux & B. Bril (Eds.), *Stone knapping: The necessary conditions for uniquely hominin behaviour* (331–340). Cambridge, UK: McDonald Institute.

Stout, D., & Chaminade, T. (2007). The evolutionary neuroscience of tool making. *Neuropsychologia, 45,* 1091–1100.

Stout, D., Toth, N., Schick, K., & Chaminade, T. (2008). Neural correlates of early stone age toolmaking: Technology, language and cognition in human evolution. *Philosophical Transactions of the Royal Society B, 363,* 1939–1949.

Stringer, C., & Andrews, P. (2005). *The complete world of human evolution.* London: Thames & Hudson.

Stringer, C., & Gamble, C. (1993). *In search of the Neanderthals.* London: Thames & Hudson.

Systma, S. E., Kelley, M. L., & Wymer, J. H. (2001). Development and initial validation of the child routines inventory. *Journal of Psychopathology and Behavioral Assessment, 23,* 241–251.

Tague, R. G., & Lovejoy, C. O. (1986). The obstetric pelvis of A.L. 288–1 (Lucy). *Journal of Human Evolution, 15,* 237–255.

Tan, J. H. W., & Vogel, C. (2008). Religion and trust: An experimental study. *Journal of Economic Psychology, 29,* 832–848.

Taylor, M. (1999). *Imaginary companions and the children who create them.* New York: Oxford University Press.

Taylor, M., Carlson, S. M., Maring, B. L., Gerow, L., & Charley, C. M. (2004). The characteristics and correlates of fantasy in school-age children: Imaginary companions, impersonations, and social understanding. *Developmental Psychology, 40,* 1173–1187.

Tetlock, P. E. (2003). Thinking the unthinkable: sacred values and taboo cognitions. *TRENDS in Cognitive Sciences, 7*, 320–324.

Tetlock, P. E., Kristel, O. V., Elson, S. B., Green, M. C., & Lerner, J. S. (2000). The psychology of the unthinkable: Taboo trade-offs, forbidden base rates, and heretical counterfactuals. *Journal of Personality and Social Psychology, 78*, 853–870.

Thieme, H. (2005). The Lower Paleolithic art of hunting: The case of Schöningen 13 II-4, Lower Saxony, Germany. In C. Gamble & M. Porr (Eds.), *The hominid individual in context: Archaeological investigations of Lower and Middle Paleolithic landscapes, locales, and artefacts* (115–132). London: Routledge.

Thomas, E. M. (1989). *The harmless people.* New York: Vintage Books.

Tinbergen, N. (1952). "Derived' activities: Their causation, biological significance, origin and emancipation during evolution. *Quarterly Biological Review, 27*, 1–32.

Tomasello, M. (2011). Human culture in evolutionary perspective. In M. Gelfand (Ed.), *Advances in culture and psychology* (5–51). Oxford: Oxford University Press.

Tomasello, M., & Carpenter, M. (2005). The emergence of social cognition in three young chimpanzees. *Monographs of the Society for Research in Child Development, 70*(1), vii–132.

Toth, N., & Schick, K. D. (2009). The Oldowan: The tool making of early hominins and chimpanzees compared. *Annual Review of Anthropology, 38*, 289–305.

Toth, N., Schick, K. D., Savage-Rumbaugh, S., Sevcik, R. A., & Rumbaugh, D. M. (1993). Pan the toolmaker: Investigations into stone tool-making and tool-using capabilities of a bonobo (*Pan paniscus*). *Journal of Archaeological Science, 20*, 81–91.

Townsend, J. B. (1999). Shamanism. In S. D. Glazier (Ed.), *Anthropology of religion* (429–469). Westport, CT: Praeger.

Tronick, E. (2003). Things still to be done on the still-face effect. *Infancy, 4*, 475–482.

Tronick, E., Als, H., & Adamson, L. (1979). The structure of face to face communicative interactions. In M. Bullowa (Ed.), *Before speech* (349–370). Cambridge: Cambridge University Press.

Tronick, E., Als, H., Adamson, L., Wise, S., & Brazelton, T. B. (1978). The infant's response to entrapment between contradictory messages in face-to-face interaction. *Journal of the American Academy of Child and Adolescent Psychiatry, 17*, 1–13.

Valdesolo, P., & DeSteno, D. (2011). Synchrony and the social tuning of compassion. *Emotion, 11*, 262–266.

Vallverdu, J., et al. (2010). Sleeping activity area within the site structure of archaic human groups: Evidence from Abric Romani Level N combustion activity areas. *Current Anthropology, 51*, 137–145.

Vandiver, P., Soffer, O., Klima, B., & Svoboda, J. (1989). The origins of ceramic technology at Dolni Vestonice, Czechoslovakia. *Science, 246*, 1004.

Van Gennep, A. (1960). *Rites of passage.* London: Routledge & Kegan Paul.

Vanhaeren, M., & d'Errico, F. (2005). Grave goods from the Saint-Germain-la-Rivière burial: Evidence for social inequality in the Upper Paleolithic. *Journal of Anthropological Archaeology, 24,* 117–134.

Vanhaeren, M., d'Errico, F., Stringer, C., James, S. L., Todd, J. A., & Mienis, H. K. (2006). Middle Paleolithic shell beads in Israel and Algeria. *Science,* 312, 1785–1788.

Vernant, J. P. (1996). The refusal of Odysseus. In S. L. Schein (Ed.), *Reading the Od-yssey: Selected interpretive essays* (185–189). Princeton, NJ: Princeton University Press.

Vitebsky, P. (2000). Shamanism. In G. Harvey (Ed.), *Indigenous religions* (55–67). London: Cassell.

Wachholtz, A. B., & Pargament, K. I. (2005). Is spirituality a critical ingredient of meditation? Comparing the effects of spiritual mediation, secular meditation, and relaxation on spiritual, psychological, cardiac, and pain outcomes. *Journal of Behavioral Medicine, 28,* 369–384.

Wadley, L., Hodgskiss, T., & Grant, M. (2009). Implications for complex cognition from the hafting of tools with compound adhesives in the Middle Stone Age, South Africa. *Proceedings of the National Academy of Sciences, 106,* 9590–9594.

Wallach, H., Newman, E. B., & Rosenzweig, M. R. (1949). The precedence effect in sound localization. *American Journal of Psychology, 62,* 315–336.

Ward, C., Kimbel, W., & Johanson, D. (2011). Complete fourth metatarsal and arches in the foot of *Australopithecus afarensis. Science, 331,* 750–753.

Warneken, F., Chen, F., & Tomasello, M. (2006). Cooperative activities in young children and chimpanzees. *Child Development, 77,* 640–663 .

Warneken, F., Hare, B., Melis, A., Hanus, D., & Tomasello, M. (2007). Spontaneous altruism by chimpanzees and young children. *PLOS Biology, 5,* e184.

Warneken, F., & Tomasello, M. (2006). Altruistic helping in human infants and young chimpanzees. *Science, 31,* 1301–1303.

Wason, P. C., & Evans, J. (1975). Dual processes in reasoning. *Cognition, 3,* 141–154.

Weber, M. (1963). *The sociology of religion.* Boston: Beacon Press.

Webster, C. (1986). Puritanism, separatism, and science. In D.C. Lindberg & R. L. Numbers (Eds.), *God and nature: Historical essays on the encounter between Christianity and Science* (192–217). Berkeley: University of California Press.

Weiss, K. M. (1981). Evolutionary perspectives on human aging. In P. Amoss & S. Harrell (eds.), *Other Ways of Growing Old* (25–28). Palo Alto, CA: Stanford University Press.

Wells, K. D. (1977). The social behavior of anuran amphibians. *Animal Behavior, 25,* 666–693.

Westergaard, G. C., Liv, C., Haynie, M. K., & Suomi, S. J. (2000). A comparative study of aimed throwing by monkeys and humans. *Neuropsychologia, 38,* 1511–1517.

Westermann, C. (1994). *Genesis 1–11*. Minneapolis: Fortress Press.

White, R. (1985). *Upper Paleolithic land use in the Perigord*. London: British Archaeological Reports.

White, R. (1993). Technological and social dimensions of "Aurignacian age" body ornaments across Europe. In H. Knecht, A. Pike-Tay, & R. White (Eds.), *Before Lascaux* (277–299). Boca Raton, FL: CRC Press.

White, R. (2003). *Prehistoric art: The symbolic journey of humankind*. New York: Abrams.

White, T., Asfaw, B., & DeGusta, D. (2003). Pleistocene *Homo sapiens* from Middle Awash, Ethiopia. *Nature, 423*, 742–747.

Whitehead, A. N. (1925/1967). *Science and the modern world*. New York: Free Press.

Whitehouse, H. (1996). Rites of terror: Emotion, metaphor and memory in Melanesian cults. *Journal of the Royal Anthropological Institute, 2*, 703–715.

Whiten, A. (1999). The evolution of deep social mind in humans. In M. C. Corballis & S. E. G. Lea (Eds.), *The descent of mind* (173–193). Oxford: Oxford University Press.

Whitham, J. C., & Maestripieri, D. (2003). Primate rituals: The function of greetings between male guinea baboons. *Ethology, 109*, 847–859.

Wiech, K., Farias, M., Kahane, G., Shackel, N., Tiede, W., & Tracey, I. (2008). An fMRI study measuring analgesia enhanced by religion as a belief system. *Pain, 139*, 467–476.

Williams, L. (1967). *The dancing chimpanzee: A study of primitive music in relation to the vocalizing and rhythmic action of apes*. New York: Norton.

Williamson, R. A., Jaswal, V. K., & Meltzoff, A. N. (2010). Learning the rules: Observation and imitation of a sorting strategy by 36-month-old children, *Developmental Psychology, 46*, 57–65.

Wiltermuth, S. S., & Heath, C. (2009). Synchrony and cooperation. *Psychological Science, 20*, 1–5.

Winkelman, M. (2002). Shamanism and cognitive evolution. *Cambridge Archaeological Journal, 12*, 71–101.

Winton, V. (2005). An investigation of knapping-skill development in the manufacture of Palaeolithic handaxes. In V. Roux & B. Bril (Eds.), *Stone knapping: The necessary conditions for uniquely hominin behaviour* (109–116). Cambridge, UK: McDonald Institute.

Wolin, S. J., & Bennett, L. A. (1984). Family rituals. *Family Process, 23*, 401–420.

Wrangham, R. W. (2009). *Catching fire: How cooking made us human*. New York: Basic Books.

Wrangham, R. W., Jones, J. H., Laden, G., Pilbeam, D., & Conklin-Brittain, N. (1999). The raw and the stolen: Cooking and the ecology of human origins. *Current Anthropology, 40*, 567–594.

Wynn, T. (2002). Archaeology and cognitive evolution. *Behavioral and Brain Sciences, 25,* 389–402.

Wynn, T., & Coolidge, F. L. (2012). *How to think like a Neandertal.* Oxford: Oxford University Press.

Wynn, T., & Tierson, F. (1990). Regional comparison of the shapes of later Acheulean Handaxes. *American Anthropologist, 92,* 73–84.

Yamagishi, T. (2007). The social exchange heuristic: A psychological mechanism that makes a system of generalized exchange self-sustaining. In M. Radford, S. Ohnuma, & T. Yamagishi (Eds.). *Cultural and ecological foundations of the mind* (11–37). Sapporo, Japan: Hokkaido University Press.

Yamagishi, T., Jin, N., & Kiyonari, T. (1999). Bounded generalized reciprocity: In-group boasting and ingroup favoritism. *Advances in Group Processes, 16,* 161–197.

Yamagishi, T., & Mifune, N. (2009). Social exchange and solidarity: In-group love or our out-group hate? *Evolution and Human Behavior, 30,* 229–237.

Yang, Z., & Schank, J. C. (2006). Women do not synchronize their menstrual cycles. *Human Nature, 17,* 434–447.

Ybarra, O., Burnstein, E., Winkielman, P., Keller, M. C., Manis, M., Chan, E., & Rodriguez, J. (2008). Mental exercising through simple socializing: Social interaction promotes general cognitive functioning. *Personality and Social Psychology Bulletin, 34,* 248–259.

Ybarra, O., Winkielman, P. Yeh, I., Burnstein, E., & Kavanagh, L. (2011). Friends (and sometimes enemies) with cognitive benefits: What types of social interactions boost executive functioning? *Social Psychological and Personality Science, 2,* 253–261.

Zeidan, F., Martucci, K. T., Kraft, R. A., et al. (2011). Brain mechanisms supporting the modulation of pain by mindfulness meditation. *Journal of Neuroscience, 31,* 5540–5548.

Zohar, A. H., & Felz, L. (2001). Ritualistic behavior in young children. *Journal of Abnormal Child Psychology, 29,* 121–128.

Index